Leading Pharmaceutical Innovation

Oliver Gassmann · Alexander Schuhmacher
Max von Zedtwitz · Gerrit Reepmeyer

Leading Pharmaceutical Innovation

How to Win the Life Science Race

Third Edition

 Springer

Oliver Gassmann
Institute of Technology Management
University of St. Gallen
St. Gallen, Switzerland

Max von Zedtwitz
Department of Strategic Management
Kaunas University of Technology
Kaunas, Lithuania

Alexander Schuhmacher
Faculty of Applied Chemistry
Reutlingen University
Reutlingen, Germany

Gerrit Reepmeyer
Novi, MI, USA

ISBN 978-3-030-09782-0   ISBN 978-3-319-66833-8   (eBook)
https://doi.org/10.1007/978-3-319-66833-8

Printed on acid-free paper

This Springer imprint is published by the registered company Springer International Publishing AG part of Springer Nature.
The registered company address is: Gewerbestrasse 11, 6330 Cham, Switzerland

Preface

The life sciences have never been an easy industry, and they are about to become even more treacherous to navigate. New technologies speed up innovation, new competitors emerge from both new geographical corners of the world and formerly distant segments at home, and customers become more sophisticated and more demanding. The healthcare sector is becoming a pressure cooker, and pharmaceutical companies are right in the middle of it.

We have been investigating the healthcare sector and leading pharmaceutical companies for more than a decade, and here are just some of the shifting patterns that we see:

- Scientists create more and more data on diseases and the human organisms, producing gene maps that pinpoint the sources of our afflictions. This is promising for the progress of personalized medicine, but it also raises ethical and moral expectations on pharma companies to protect and commercialize this knowledge. Treating a patient is increasingly an economic rather than a medical decision.
- Pharma innovation does not end with the conclusion of clinical trials. To be successful, pharma companies must be able to maneuver public opinion, regulatory decisions, and financial markets alike.
- Emerging countries are catching up: China, a relative backwater for pharma until just a few years ago, now makes an impact on the global life science scene. Global pharma firms set up research centers in Shanghai, Beijing, and other Chinese cities. Chinese research institutes make key contributions to pharmaceutical research, even winning Nobel prizes. No other country has ever attempted a healthcare reform affecting more than a billion people, but China is doing so as you are reading this book. In China, the numbers are always staggering: three hundred thousand hospitals and healthcare facilities offer their services to patients, and 6800 pharmaceutical companies fight for market share.
- Digital health will reshape the pharmaceutical sector dramatically within the next years. Mobile technologies and smart devices developed in the consumer electronic industry enter the healthcare markets. Most pharma companies neglect the disruption from the low end. But as quality of those new entrants is improving, a growing share of the healthcare budget will be reallocated. Digitalization will

change the roles of patients, consumers, and financing partners. Data becomes the new currency, and new business models are arising already. The regulator is challenged in many regions already.

- New technologies such as nanotechnologies, blockchain, deep learning, and artificial intelligence will become central for innovation in the healthcare sector, but most pharmaceutical firms today are not familiar with them.
- Being more open in innovation has been on the agenda of most of the big pharma companies for some years now. Cost pressures, growth expectations, globalization, the growing complexity of R&D, as well as the increasing power of technology providers force pharma executives to redesign their R&D departments into more trim, more flexible, and more open organizations. Technological fields raise the bar for everyone, and even the large pharma companies cannot finance and discover them alone. Collaboration is the new imperative in the industry.

Many challenges remain: the most important one is the widening innovation productivity gap in the pharma industry. We have spared no effort to analyze how industry leaders face these challenges, what tools they deploy, and what new solutions they are testing.

Specifically, we have updated this third edition to include the latest technology and industry information, and we have paid particular attention to new management models such as open innovation, systematic partnering, outlicensing, and international diversification of R&D. This leads not just to new forms of collaboration in the pharmaceutical industry but also to new business models. We introduce a more balanced global perspective by reducing the focus on Switzerland as a lead country and adding more industry cases from the USA and Asia. All in all, this third edition is quite a different book from the earlier two versions some ten years ago.

In this third edition, we revised the book completely and expanded it as follows:

- Chapter 1 comprises up-to-date information on the pharma innovation arena, ranging from the importance of innovation in the pharmaceutical sector and the relevance of pharma companies as key investors in R&D worldwide to the role of blockbusters and mergers and acquisitions as key growth factors for the industry.
- Chapter 2 offers more insight on the complexity of the industry by describing the industry's classification, its product groups, and the six driving forces affecting any company in the pharmaceutical industry: the bargaining power of suppliers and buyers, the risk of entry of new competitors, the uncertainty related with emergence of substitution products (e.g., generics, medical devices, or alternative therapies), the rivalry among established companies, and the power of regulators.
- Chapter 3 gives an update on biologics and the increased role of biotech companies as technology providers for the whole sector. New sciences and technologies are pushing the industry as the underlying drivers for innovation. In this context, we provide the latest information on how drug discovery is performed today—from target identification and validation to the roles of HTS, bioinformatics and 'omics technologies, big data, pharmacogenetics and pharmacogenomics, personalized medicine, and computer-based drug discovery.

- We revised Chap. 4 completely to highlight the relevance of pipeline management, given the changes in the regulatory environment, the huge investments needed in pharmaceutical R&D, and the low overall success rates of drug R&D. We also provide a detailed description of the standard R&D process with a case in point on the importance of preclinical safety studies. Best-in-class project and portfolio management is a must for successful R&D today—including an understanding of why to evaluate a drug project financially, how to analyze a project portfolio, and what needs to be done to manage a project pipeline.

- The changes we made in Chap. 5 reflect the trend that pharma firms increasingly concentrate on their core advantages and try to involve outside innovators. Outsourcing, in- and outlicensing, research collaborations, partnerships, and pharma–pharma co-developments are becoming must-have tools in pharmaceutical R&D. Outsourcing plays a key role in innovation management today and is a core element to access external technologies or to reduce costs of R&D. Here, we highlight the general conditions for outsourcing in pharma and biotech scenarios, exemplified by two strong cases. CROs have also assumed a more prominent role in the pharma R&D value chain, although perhaps not yet as important as the subcontractors in the automobile sector, but with an increasing relevance for all kinds of healthcare firms. Pfizer and its CRO partners provide one example of the strategic role of contract research organizations for big pharma to manage the R&D pipeline flexibly. Takeda's collaboration with the Center for iPS Cell Research Application (CiRA) at Kyoto University shows how pharma companies need to collaborate with world leading competence centers to access world-class science. Next, the R&D pipeline can also be enhanced by licensing drug candidates along the entire value chain—and not only in the late phases of drug development. Co-development agreements between pharma partners such as Morphosys/Novartis, Pfizer/Merck, and Astra Zeneca/Innate Pharma complement the complex models of accessing and managing innovation.

- The new Chap. 6 on open innovation covers one of the most exciting trends of the last decade. We provide an overview on different open innovation models—virtual R&D, crowdsourcing, open source, private–public partnerships, innovation centers, and venture funds—alongside with several pharma cases. Crowdsourcing is illustrated with examples from YourEncore, Innocentive, the European Lead Factory, and Grants4Targets. Venture funds as a tool to mitigate risk while accessing high-risk early pipeline project and disruptive technologies are illustrated by the Novartis Venture Fund and the Boehringer Ingelheim Venture Fund.

- Chapter 7 addresses the internationalization of R&D, a change from large domestic R&D sites to a network of international, multilocation R&D. Today, pharma companies run a mixed model of in-house R&D with a global footprint combined with open innovation partnerships to access innovation globally. The reasons for this trend are obvious, such as reduced R&D costs, talent acquisition, and technology-friendly legal conditions. In the last years, the emerging markets in China and India have become the fourth global center of gravity for pharmaceutical R&D, supplementing the traditional pharma centers in the USA, Europe, and Japan. Especially China has made quite some impact in this context, both as a market and as a center for drug development, on the back of an impressive overall economic resurgence.

- Finally, we outline our views on future trends and directions by reviewing them at three levels: R&D efficiency and effectiveness, new business models, and leading people and change. This condensed set of future directions is our final chapter of the book and hopefully stimulates new ideas and new leads for management innovation.

We hope that this book contributes to a better understanding of one of the most fascinating industries and that it helps its leaders to increase competitiveness of their companies.

Whether you are a pharma executive, an innovation researcher, a medical professional, or a member of the general public genuinely interested in the pharmaceutical industry, we hope that you will join us in our appreciation for the tremendous work that is being done by the millions of people engaged in this industry with the respectable goal to provide mankind with better drugs and therapies to live and survive.

St. Gallen, Switzerland   Oliver Gassmann
Reutlingen, Germany   Alexander Schuhmacher
Kaunas, Lithuania   Max von Zedtwitz
Novi, MI   Gerrit Reepmeyer
June 2017

Testimonials

"Health care innovation is possibly the greatest opportunity and challenge of our generation. This important book shows how pharmaceutical companies will continue to play an instrument role in making our lives better."
Stefan Thomke, William Barclay Harding Professor of Business Administration, Harvard Business School

"Very comprehensive review and analysis of current challenges for the biopharmaceutical industry. To stay competitive in this new technology driven environment, the industry started to develop new partnership models to close the innovation gap and provide patients with the relevant healthcare toolkits at the convergence of pharmaceuticals, diagnostics and IT technologies. Very exciting times ahead well depicted in this book!"
Karima Boubekeur, VP Emerging Portfolio and Search & Evaluation, AstraZeneca

"Great. The 3rd edition is not only an update. It's an outstandingly featured summary on the challenges of pharma innovation."
Eckard von Keutz, SVP and Head of Early Development, Bayer Healthcare

". . .a profoundly researched and comprehensively illustrated creation of the value-drivers of pharmaceutical innovation."
Florian Gantner, VP Translational Medicine & Clinical Pharmacology, Boehringer Ingelheim

"*Leading Pharmaceutical Innovation*' is the benchmark on pharmaceutical innovation management."
Alexander Musil, CFO Takeda Mexico

"Globalization, internationalization and open innovation are three major trends in the industry. Gassmann, Schuhmacher, von Zedtwitz and Reepmeyer provide most valuable insights in these topics. I highly recommend reading this book."
Mathias Schmidt, CEO Armagen

"I recommend this new book to pharma managers and researchers in this industry."
Maximilian von Wuelfing, General Manager Mylan BeLux

"This text is an excellent information source for innovation concepts in the modern era when pharma productivity is getting ever challenging in this global environment"
Praveen Tyle, Executive Vice President R&D, Lexicon Pharmaceuticals

"The book remains the key important publication of pharmaceutical innovation for business executives, scientists and students. The 3rd edition was updated especially regarding innovation alliances and addresses the key challenges that pharmaceutical R&D is facing nowadays."
Katharina Caspary, Director Pharmacovigilance, Horizon Pharma

"Excellent book for strategies on the management of pharma innovation. Highly recommendable for creating the bigger picture and new perspectives for action."
Ingo Gaida, Head of IT Operations R&D, Bayer Business Services

"Analytical & critical - a "spot on" review of the constantly changing pharma environment."
Guenther Forster, former SVP Medical Development & Strategies, MerckSerono

"This great book helps to understand the complexity of the challenges the pharmaceutical industry is facing."
Ingo Henes, SVP Human Relations at Rentschler Biotechnologie

Contents

About the Authors

Oliver Gassmann is professor of technology management at the University of St. Gallen (Switzerland) and managing director of the Institute of Technology Management. After completing his PhD in 1996, he was head of corporate research at Schindler. He published over 400 papers in leading international management journals. In 2014, he was awarded the Scholary Impact Award by the Journal of Management and in 2015 with the Citation Excellence Award of the Emerald Group. He also serves as a founder and advisor on several boards.

Alexander Schuhmacher is professor at Reutlingen University (Germany). He teaches in business management and medical science. He officiates as senator of Reutlingen University, and he also serves as managing director of the Knowledge Foundation @ Reutlingen University. He worked for 14 years in various R&D positions in the pharmaceutical industry. As member of Nycomed's senior R&D management team, he headed the Strategic Planning and Business Support function and coheaded the R&D integration team following the merger of Nycomed and ALTANA Pharma.

Max von Zedtwitz is professor at Kaunas University of Technology (Lithuania) and director of the GLORAD Center for Global R&D and Innovation. Previously, he was a professor at Tsinghua, Peking University, Skoltech, and IMD and a vice president at PRTM. In 2009, he was recognized by IAMOT as one of the fifty most influential innovation scholars worldwide. A frequent public speaker, he has appeared on television and has been cited in the Economist, China Daily, the South China Morning Post, and the New York Times.

Gerrit Reepmeyer is cofounder & COO of Guardhat Inc., a start-up based in Detroit (USA). Prior to that, he held various executive leadership positions in industrial and technology companies. He started his career in management consulting with McKinsey & Company. He has a PhD degree from the University of St. Gallen and was a research scholar at Columbia University in New York. He has coauthored several books and publications on technology management, and he is co-inventor of five patents.

Innovation: Key to Success in the Pharmaceutical Industry

1

". . . lasting innovation is our biggest gift to society."
Dr. Severin Schwan,
CEO Roche

1.1 The Productivity Paradox

Despite its high research and development (R&D) intensity, the pharmaceutical industry is facing an increasingly challenging situation. On average, only 1 out of 10,000 substances tested preclinically becomes a marketed product. And only three out of ten drugs generate revenues that meet or exceed the average R&D costs (Grabowski et al. 2002).

By definition, R&D efficiency is the ratio of (financial) input in R&D versus its (nominal) output. The black-box in between consists of the R&D pipeline, screening and other drug discovery technologies, worldwide collaboration networks in pre-clinical research and drug development, and an armada of licensing and partnering agreements with universities, competitors and biotechnology start-ups. Still, R&D performance of the major research-based pharmaceutical companies is sub-optimal:

- The overall R&D pipeline output is low;
- Costs of R&D are enormous, driven by larger and more complex clinical studies and expensive drug discovery technologies;
- Over-supply of 'me-too' launches and a lack of genuinely innovative drugs make it difficult to replace revenues last after patent expiration;
- Longlasting, protracted and complex clinical trials and administrative procedures reduce the marketed shelf life of patented products.

As indicated by member companies of the Pharmaceutical Research and Manufacturers of America (PhRMA), the total R&D epxenditure of U.S. based pharmaceutical companies increased from US$15.2 billion (1995) to US$51.2

© Springer International Publishing AG, part of Springer Nature 2018
O. Gassmann et al., *Leading Pharmaceutical Innovation*,
https://doi.org/10.1007/978-3-319-66833-8_1

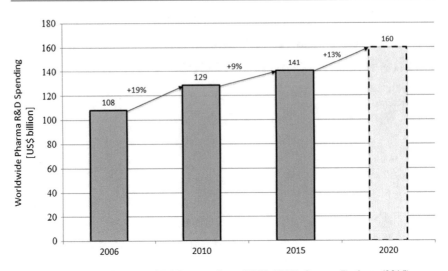

Fig. 1.1 Pharma worldwide total R&D expenditure (2006–2020). Source: Evaluate (2015)

billion in 2014 (PhRMA 2015). Worldwide, the total pharmaceutical and biotechnology R&D spending rose from US$108 billion (2006) and US$129 billion (2010) to US$141 billion in 2015 (CAGR 2006–2015: +3.1%) (see Fig. 1.1).

The pharmaceutical sector is a top investor in global R&D. According to the 2015 EU Industrial R&D Investment Scoreboard report, the 15 largest pharmaceutical companies are in the top-50 league of global innovators and R&D investors for the fiscal year of 2014/2015 (see Fig. 1.2). For example, Novartis (8.217 billion euros) outspent the makers of popular brands such as Google (8.098 billion euros), Toyota Motors (6.858 billion euros), General Motors (6.095 billion euros) or Apple (4.976 billion euros) (European Commission 2015).

Pharmaceutical companies invest one of the worldwide highest shares of total sales back into R&D (see Table 1.1, Fig. 1.3). On average, companies of the pharmaceutical industry invested 14.4% of their total sales in R&D—a significant higher proportion compared to other sectors, such as software and computer services (10.4%), automobiles and parts (4.3%) or chemicals (2.6%) (European Commission 2014). The level of R&D investment in the pharmaceutical sector is driven by the complexity of the underlying R&D involved, the low success rates of taking drug candidates from the early research stages all the way to the end, the technology and resource-intensivity of the scientific investigations, and the need to develop drugs for a global market right away at product launch.

The enormous R&D expenditures per company result from the sum of high direct costs per R&D projects that need to be borne to bring one single New Molecular Entity (NME) successfully to its market launch (Paul et al. 2010). We must also remember that there is an unproportionally large number of failed R&D projects whose results never see the light of the day, but for which the investing company needs to pay regardless. At best, a company can hope to sell intermediary results or

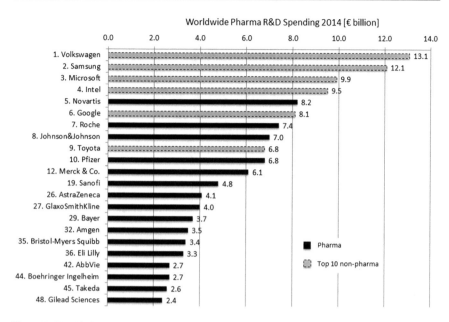

Fig. 1.2 Worldwide top R&D investors in 2014. Source: European Commission (2015)

Table 1.1 The largest R&D spenders in the pharmaceutical industry (for 2014, in US$ billion)

#	Company	Origin	R&D spending	Total revenues
1	Novartis	CH	9.3	46.1
2	Roche	CH	8.6	40.1
3	Pfizer	US	7.2	44.5
4	Merck & Co.	US	6.5	36.6
5	Johnson & Johnson	US	6.2	30.7
6	Sanofi	FR	6.2	38.2
7	GlaxoSmithKline	UK	5.1	30.3
8	AstraZeneca	UK	4.9	25.7
9	Eli Lilly	US	4.4	16.3
10	Amgen	US	4.1	19.3

Source: 2015 Annual reports of companies listed

byproducts in the open market, or perhaps license off some more advanced compounds to better suited companies.

Unlike in most other industries, R&D expenditures make up a large share of the cost structure of a newly developed drug. They represent the 20–40% contribution to the overall costs of a newly developed drug (Table 1.2).

The generally science-driven linear representation of pharmaceutical innovation allows to allocate costs into fairly easy to describe phases in the pharmaceutical

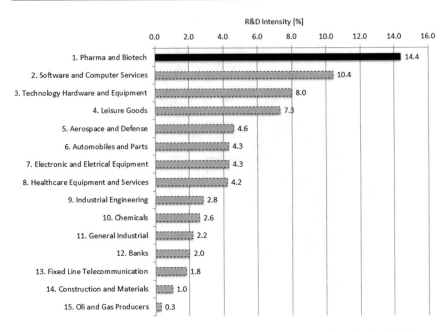

Fig. 1.3 Overall R&D intensity by industrial sector. Source: European Commission (2014)

Table 1.2 Average cost structure of a new developed drug.

Relative contribution (%)	Cost factors
20–40	Research, Development, Licenses
15–30	Production
5–15	Technical and Administrative Costs
20–30	Marketing and Distribution
20–35	Margin

Source: Pharma Information (2002)

R&D process (see Fig. 1.4). It is exactly because pharmaceutical R&D is mainly science-driven and integrated in a strictly regulated healthcare environment that it can be captured and described so well by a linear innovation model.

Science-based R&D is characterized by a fairly high propensity of failure throughout the entire R&D process. This failure-rate, also called "project risk" or "attrition rate", is inherent to pharmaceutical innovation; we will address it further below. In its ultimate consequence, it means that much of the money spend on R&D projects in the early phases will not result in a marketable product—the problem is that nobody knows which project will succeed in the end.

The long timelines of pharmaceutical R&D also have a negative effect on R&D costs: In addition to the already very time-consuming activities in applied research and preclinical/clinical development, the regulatory review and approval process further extends time-to-market for R&D. Although the U.S. Food and Drug

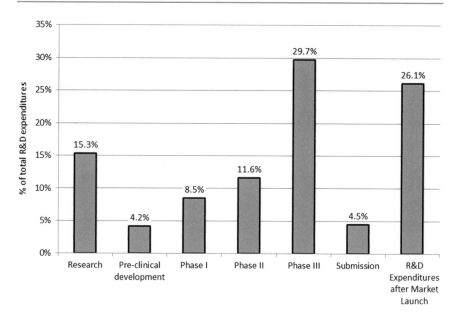

Fig. 1.4 Proportion of total R&D expenditures by phase. Source: CMR Int'l data from clarivate. com website

Administration (FDA) has reduced its processing time since the Prescription Drug User Fee Act (PDUFA) came into force in 1992, the target duration for a standard review still is 10 months. All in all, the average time for a drug project to pass through preclinical and clinical development is approximately 9 years (see Fig. 1.5), and the total time-to-market including drug discovery can easily amount to 14 years.

These long timelines typical for pharmaceutical R&D induce further challenges. First, expenditures for R&D projects start early in drug discovery and need to be capitalized for years till the date of return-on-investment (ROI) of the marketed drug. This effect results in an enormous increase in the overall R&D expenditures and can double the costs per drug project up to reported US$1.8 billion per NME (Paul et al. 2010).

Second, many pharmaceutical companies follow similar R&D concepts resulting from given market needs. They address the same diseases by identical biological mechanisms and drug targets and aim to provide new medicaments to treat the patients suffering from the same diseases. The long drug R&D timelines increase the risk of competition and reduce the market potential and success of new drugs.

Third, the date of patent expiration (given after a fixed number of years after the patent was granted) and thus the almost inevitable arrival of generic competition, influences the ROI of drugs. Any delay in drug discovery or development impacts the commercially usable patent term negatively.

The concrete problem for the pharmaceutical industry results from contrasting the output of pharmaceutical R&D (the number of NMEs launched to the markets, see Fig. 1.6) to the input (the total costs of R&D). At the company-level, the so-called

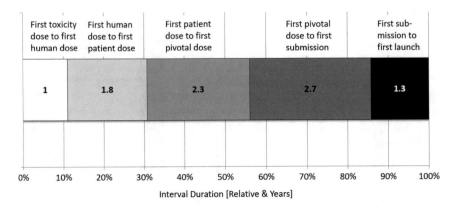

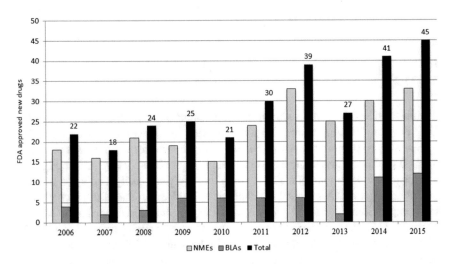

Fig. 1.5 Absolute (in years) and relative (in %) composite median interval duration (2009–2013). Source: CMR Int'l data from clarivate.com website

Fig. 1.6 FDA drug approvals 2006–2015. Source: Mullard (2016)

R&D efficiency of some of the key players in the industry is so low that they need to invest more than US$3 billion per new drug launched (Schuhmacher et al. 2016). In the wider context of the pharma sector, the input/output-ratio split over the past 60 years shows that to get one NME approved one had to double R&D investments every 9 years (Scannell et al. 2012). The resulting productivity gap is illustrated in Fig. 1.7.

The reasons for the low R&D efficiency of the industry are:

- The low success rates of R&D projects;
- The capitalization of R&D costs over the lengthy time period of drug R&D;

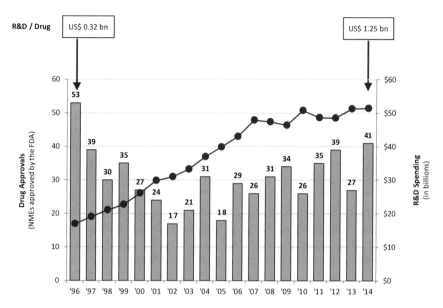

Fig. 1.7 The productivity gap in pharmaceutical research and development. Source: Data from Pharma.org and Fda.gov websites

- The insufficient numbers of projects in early R&D phases;
- The lower risk tolerance of both regulators and the society as a whole contributes negatively to NME approval rates and the development-associated costs;
- New medications are based on technically more complex investigations;
- The increasing number of approved drugs raises the hurdle for approval and reimbursement of NMEs;
- The lower number of well-established pharmaceutical companies and, for this reason, the reduced number of investors that accept the risk of pharmaceutical R&D;
- The increasing number of mergers & acquisitions (M&As) that influences the productivity of R&D organizations negatively;
- The time delays caused by licensing, co-development, and joint venture negotiations.

A significant increase in productivity in pharmaceutical innovation is needed in order to close the widening productivity gap and to meet the high revenue growth expectations of the industry and investors.

1.2   The Blockbuster Imperative

Blockbuster drugs—drugs with at least US$1 billion in annual revenues—are still a growth driver for most leading pharmaceutical companies and are often quoted as the only viable way to meet high growth expectations. One reason of the industry's reliance on the blockbuster strategy is that blockbuster drugs offer relatively high returns compared to lower value drugs, again a consequence of the substantial risks, time lags and costs involved in product development and commercialization.

In 2014, 64 ethical pharmaceutical products were considered blockbuster drugs. The best-selling drug in 2014 was Humira[1] (Adalimumab) from Abbvie with world-wide prescription sales of US$12.890 billion, followed by Solvadi (US$10.283 billion, Gilead Sciences) and Enbrel (US$8.915 billion, Amgen/Pfizer/Takeda). In the past, blockbuster drugs were high-volume drugs addressing needs of many patients, such as in gastrenetrology or respiratory diseases. Nowadays, specialty medicines, such as Solvadi or the cancer drugs, also have the potential to become blockbusters because of the high prizes realized for these types of medicine. For example, 3.2 million U.S. Americans suffer from Hepatitis C. Solvadi, the Hepatitis C drug of **Gilead Sciences**, is forecasted to capture 50% of the entire anti-viral segment by 2020 (Evaluate 2015). Today, it sold in the U.S. at a prize of US$84,000 for one course of treatment, giving the drug a lifetime market potential in the U.S. of more than US$130 billion. Assuming the drug is sold in middle-income countries such as Argentina, Brazil, China, and Russia at a prize of less than US$10,000, it could generate additional sales of more than US$250 billion.

However, dependency on blockbuster revenues can cause serious challenges when facing patent expiration. The emergence and growing power of generic companies, such as Teva Pharmaceuticals or Sandoz, pose a growing threat to established pharmaceutical companies. Some ethical drugs might lose up to 80% in market share within just one quarter of a year after patent expiration, exposing several US$ billions worth in revenues to generic competition. For instance, in 2011 **Pfizer** had to face loss of exclusivity (LOE) of its best-selling drug Lipitor (Atorvastatin), a cholesterol lowering drug. It eventually lost 59% of its worldwide sales and 81% of its U.S. revenues when sales collapsed from US$9.6 billion in 2011 to US$3.9 billion in 2012. Pfizer anticpated this loss-of-sale and took this opportunity to update its R&D pipeline, cut its R&D costs, and search for other growth options elsewhere (such as the attractive vaccine business of Wyeth). As a matter of fact, Pfizer acquired Wyeth for US$68 billion in 2009 and reduced its annual R&D expenditures from US$9.4 billion in 2010 to US$6.7 billion in 2013.

The much-discussed patent cliff was in the years of 2011–2015, with an alltime high in sales losses by generic competition in 2012 of US$37 billion. The expected sales losses by patent expiration is likely to decline to a reduced level of US$12–18

---

[1]As is customary, all names of medicines, drugs and product labels should be considered as trademarked. We have no commercial interest in any of the companies or their products mentioned in the book.

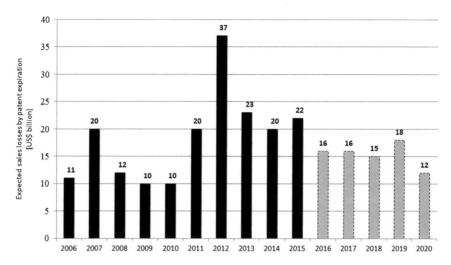

Fig. 1.8 Worldwide expected sales losses by patent expiration 2006–2020. Source: Evaluate (2015)

billion annually in the next years (see Fig. 1.8). It looks as if the industry is back on track and most of the leading pharmaceutical companies commercialize more block-buster drugs today than in the past (Table 1.3). In particular, the trend toward biotherapeutics (biologics) offer benefits, as they have revolutionzed treatment options in some diseases, and they offer better patent protection. Furthermore, the market penetration of biosimilars (biogenerics) is very low so far: only two biosimilars have been approved in the U.S. by 2015.

Table 1.4 lists the ten best selling drugs of 2014 including the therapeutic area, the indication, the marketing company and its worldwide sales. Humira (Adalimumab, US$12.5 billion) and Solvadi (Sofosbuvir, US$10.2 billion) are both mega blockbusters, with possible aspirations to become the best-selling drug of all time, Pfizer's Lipitor, with its global sales of US$13.7 billion in 2006. All of these top-10 drugs are in the top list of pharma's biggest blockbusters ever. Four of them are conventional drugs and the other six are biologics; together they accounted for worldwide sales of US$76.118 billion in 2014.

After US$542 billion in 2006 and US$687 billion in 2010, the pharmaceutical market grew to a level of US$723 billion worldwide prescription drug sales in 2014 (see Fig. 1.9). In 2014, the global ranking of pharmaceutical companies by world-wide prescription drug sales was let by the Swiss pharma giant Novartis with global drug sales of US$46.0 billion and a market share 6.2%. Pfizer (US) and Roche (CH) are following with global sales of US$44.5 billion (6.0%) and US$40.1 billion (5.4%). Sanofi, Merck & Co., Johnson & Johnson (J&J), GlaxoSmithKline, AstraZeneca, Gilead Sciences and AbbVie complete the list of top 10 leaders of the industry (Table 1.5). However, because of the high fragmentation of the pharma-ceutical industry, they still account for less than 46% of the global market.

Table 1.3 Total number of blockbuster drugs of leading pharmaceutical companies (2005–2014)

	2005	2006	2007	2008	2009	2010	2011	2012	2013	2014	Sum
Pfizer	8	7	5	4	4	5	5	4	3	3	48
Merck	3	5	4	3	4	2	5	4	3	3	36
AstraZeneca	3	4	3	3	3	3	5	4	3	3	34
GSK	4	5	5	3	2	1	1	2	3	3	29
Eli Lilly	1	2	2	2	2	2	3	2	3	3	22
Novartis	1	1	2	2	2	2	3	3	3	3	22
BMS	3	2	2	2	3	3	1	1	1	3	21
Sanofi		1	1	1	2	1	3	3	3	3	18
Takeda	2	2	2	2	2	1	1	1	1	1	15
Boehringer			1	2	2	1	2	2	1	3	14
Amgen	1	1					4	3	1	3	13
Abbott/ Abbvie			1		1	1	2	2	2	1	10
J&J	1	1	1						3	3	9
Gilead							2	2	1	3	8
Roche									3	3	6
Astellas									1	3	4
Bayer									1	3	4
Sum	27	31	29	24	27	22	37	33	36	47	313

Source: Company Annual Reports (multiple years)

Acquisitions in the pharma industry should therefore be expected, but they are not always easy: In April 2016 **Pfizer** announced its intention to acquire Allergan for US$160 billion, essentially becoming the world leading pharmaceutical company again with estimated total revenues of approximately US$74 billion. But the deal was stopped because of political interventions in view of a possible move of the headquaters of the new Pfizer from the U.S. to Ireland and the related tax losses for the U.S.

Reviewing the pharmaceutical market in more detail, oncology was the largest therapeutic area in 2014 with worldwide sales of US$79.2 billion, followed by anti-rheumatics (US$48.8 billion) and anti-viral drugs (US$43.1 billion) (Evaluate 2015). Market growth in 2013 and 2014 came primarily from chronic disease areas, such as cancer (+8%), anti-rheumatics (+8%), anti-diabetics (+8%) and multiple sclerosis (MS, +20%) therapies, and antiviral drugs (+55%), exemplified by new breakthrough therapies, such as Gilead's Solvadis and J&J's Olysio. The anti-hypertensives (US$30.5 billion, −9%) and anti-hyperlipidaemics (US$11.7 billion, −11%) markets became less important, while the bronchodilator market (US$32.5 billion, not much growth) remains important (see Table 1.6).

Table 1.4   Global blockbuster drugs in 2014

#	Brand name (INN)	Marketing company	Therapeutic area	Primary indication	Sales in US$ million
1	Humira (Adalimumab)	AbbVie	Inflammatory disease	Rheumatoid arthritis	12.543
2	Sovaldi (Sofosbuvir)	Gilead Sciences	Gastroenterology	Hepatitis C	10.283
3	Lantus (Insulin glargine)	Sanofi	Diabetes	Diabetes	8.428
4	Rituxan (Rituximab)	Roche	Hematology	Lymphoma, leukemia	7.547
5	Avastin (Bevacizumab)	Roche	Oncolgy	Cancer	7.018
6	Seretide (Advair)	Glaxo-SmithKline	Respiratory	Asthma, COPD	6.966
7	Herceptin (Trastuzumab)	Roche	Oncology	Cancer	6.863
8	Remicade (Infliximab)	Johnson & Johnson	Inflammatory disease	Rheumatoid arthritis	5.790
9	Crestor (Rosuvastatin)	AstraZeneca	Cardiovascular	High cholesterol	5.512
10	Lyrica (Pregabalin)	Pfizer	Central nervous system	Fibromyalgia, chronic pain	5.168

INN investigational non-proprietary name; *COPD* chronic obstructive pulmonary disease
Source: Company Annual Reports

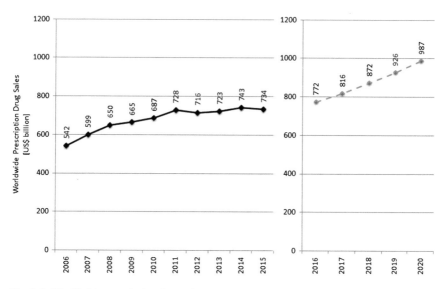

Fig. 1.9   Worldwide prescription drug sales (2006–2020). Source: Evaluate (2015)

Table 1.5 Top 10 pharmaceutical companies by worldwide prescription drug sales in 2014

#	Company	Origin	Market share (%)	Rx sales (US$ billion)
1	Novartis	CH	6.2	46.1
2	Pfizer	US	6.0	44.5
3	Roche	CH	5.4	40.1
4	Sanofi	FR	5.1	38.2
5	Merck & Co.	US	4.9	36.6
6	Johnson & Johnson	US	4.1	30.7
7	GlaxoSmithKline	UK	4.1	30.3
8	AstraZeneca	UK	3.5	25.7
9	Gilead Sciences	US	3.3	24.5
10	Abbvie	US	2.7	19.9

Rx prescription drug
Source: 2015 Annual reports of companies listed

Entering the Market Quickly

The growth rate and market share achieved in the first year after a drug's launch largely determine overall lifetime sales that can subsequently be achieved for a new product. Time-to-market is extremely important in breakthrough pharmaceuticals. The first drug in a new market captures between 40% and 60% of market share, and the second only around 15%. Coming in third already means negative business. Delaying market introduction of a blockbuster drug by only 2 months not increases the risk that a competitor seizes significant market share, it also means an estimated net loss of US$100 million, or almost US$2 million a day. Consequently, the first year of a drug's marketed life attracts the majority of promotional resources relative to any other year in the lifecycle.

This is a well-established pattern for blockbuster products (a well researched example is Pfizer's Lipitor). All of these products experienced above average sales growth in their first year on the market. Only significant external events, for example the discovery of major negative side-effects, can bring down promising new drugs that had a great first year market performance.

The market dynamics during the product launch are characterized by three closely-linked determinants. To improve the probability of a new drug to succeed, the product should be (see Reuters 2003a):

- Early to enter a particular therapy area or product class;
- Positioned relative to existing competition;
- Accompanied by heightened pre-launch awareness.

Notably, pre-launch promotion has become more important. A new product's rate of acceptance can be significantly boosted if the market is well prepared for it. The key focus of such investments is raising awareness among physicians and, eventually, patients. This is particularly important in new areas when a product is first to

Table 1.6 Worldwide prescription drug & OTC sales by therapeutic area in 2020

Therapy area	WW sales in 2014 (US$ billion)	WW sales in 2020 (US$ billion)	CAGR % growth	WW market share in 2014 (%)	WW market share in 2020 (%)
Oncology	79.2	153.1	11.6	10.1	14.9
Anti-diabetics	41.6	60.5	6.4	5.9	5.9
Anti-rheumatics	48.8	53.2	1.5	5.2	5.2
Anti-virals	43.1	49.6	2.3	4.8	4.8
Vaccines	26.7	34.7	4.4	3.4	3.4
Bronchodilator	32.5	32.5	–	3.2	3.2
Sensory organs	18.6	30.4	8.5	3.0	3.0
Anti-hypertensives	30.5	25.8	−2.8	2.5	2.5
MS-therapy	19.4	23.1	2.9	2.2	2.2
Immunosuppressants	9.2	18.6	12.5	1.8	1.8

WW worldwide, *CAGR* compound annual growth rate, *MS* multiple sclerosis
Source: Evaluate (2015)

market or if there is little awareness of the disease, its symptoms and treatment options.

Marketing departments are working closer and closer with their R&D counterparts to ensure that clinical trials are designed to meet specific market needs, and that customer value propositions are conveyed to physicians prior to launch. To this end, developing and nurturing relationships with physician opinion leaders throughout the R&D process is critical. This also helps determine unmet market needs, clinical trial design and product positioning. Good relationships with physicians drive product uptake upon launch, as opinion leaders will already be familiar with the product and its benefits.

Pre-launch marketing activities often include establishing advisory boards and sponsoring pre-launch conferences for the dissemeniation of clinical trial results to the interested medical community. They also include the direct involvement of leading physicians and medical establishments in clinical trials. By convincing opinion leaders of a drug's benefits, acceptance among late adopters can be accelerated. Early product branding further raises awareness among physicians prior to launch, increasing the likelihood of higher levels of initial uptake.

Raising pre-launch awareness also ensures that the needs of all stakeholders in the prescribing process are addressed before a product is launched. The payers' needs should also be considered as they have the final word on price and reimbursement levels.

Even though pharmaceutical companies do not necessarily need to raise awareness of a product with payers in the same way as with physicians and patients, pre-launch preparation should include health economics and outcomes studies (e.g., cost-effectiveness studies) to demonstrate product value to payers. This is particularly important in publicly funded healthcare systems operating under extensive cost containment policies.

1.3   High Risks in Drug Development

Besides its dependency on blockbuster products, the pharmaceutical industry is characterized by two other unique circumstances: its high investments in R&D and, for this reason, its strict R&D pipeline and portfolio management (see Chap. 4).

Generally speaking, innovation risks can be grouped into technical, market and commercial risks. In the pharma industy, the technical risks comprise all technical and scientific uncertainties related with drug discovery and development. The market risks relate to the question whether there are enough patients with an unmet medical need that can be treated with the new drug once launched in the market. The commercial risks depend on the uncertainties of pricing and reimbursement.

In most industries, the decision to abandon a R&D project is primarily made on the basis of economic/financial considerations related with market and commercial risks. In the pharmaceutical industry, however, most R&D projects are terminated for scientific and technical reasons, such as a lack of efficacy or insufficient safety. These risks may only become apparent late in the clinical stages when most of the investments have alredy been allocated. The failure of a drug candidate this late in the R&D process causes significant losses, as little of the product, technology or science thus far developed can be reused elsewhere (which is not the case for most other industries).

In fact, adverse news about a new compound in development can often cause a significant drop in share prices of pharmaceutical companies, destroying several US$ billion in shareholder value within minutes. For example, the U.S. biotechnology company Chimerix lost 82% of its shareholder value in 2015 after announcing that its phase III drug candidate Brincidofovir failed to prevent reactivation of cytomegalovirus (CMV). Similarly, US$12 billion in shareholder value were destroyed when Roche announced that two drug candidates for the treatment of Alzheimer's disease and cancer had failed in clinical development in 2014.

The bad news is that the scientific and technical risks of pharmaceutical R&D are still extremely high. Based on its 2014 Global R&D Performance Metrics Program, the Center for Medicine Research International (CMR) reported an overall success rate of 4.9% from first toxicity dose to market approval (see Fig. 1.10). Almost similar results were published by Paul et al. (2010) with an overall probability of technical and regulatory success (PTRS) for drug R&D of 4.1%.

Lack of efficacy, safety concerns, strategy changes, commercial reasons and operational challenges are the main causes why drugs fail in clinical development. Generally, the underlying reasons of the high attrition rates in the clinical phases can be summarized as:

- Adverse pharmacokinetics and bioavailability;
- Poor predictive preclinical animal models;
- The concept of target-based drug discovery;
- The increased complexity of clinical trials;

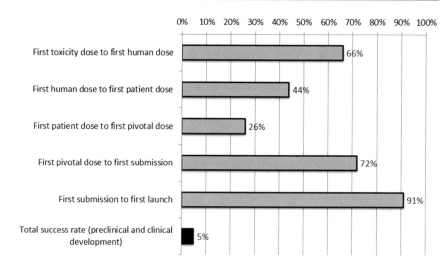

Fig. 1.10 Between phase success rates (for R&D projects entering the phase between 2008 and 2010). Source: CMR Int'l data from clarivate.com website

- The commercial demands of the marketing departments;
- The lack of know-how in smaller organizations, such as biotechnology companies.

Differentiation via Clinical Profiles

The most sustainable approach to create value in pharmaceutical innovation seems to be simple: offer a drug with a superior clinical profile compared to competitors' drugs. A drug's clinical profile is regarded as one of the most influential predictors of commercial success and consists of four criteria:

- Efficacy
- Safety/side effects
- Dosage/administration, and
- Costs

In other words, if a product is efficacious, has negligible side effects and can be administered with a convenient dosing mechanism, it is in a good position to compete in most markets.

The degree to which a product can be differentiated by any or all of these criteria varies by therapeutic market and competitive environment. For example, late market entrants offering only marginal improvements in efficacy may need to enhance their commercial prospects by competing on a low price. Alternatively, trials can be designed to target areas of unmet need, for example, efficacy in specific patient subpopulations or improved dosage schedules.

Clinical trial data are typically generated during the drug's development in the clinical phases I–III. As the pressure for product differentiation has significantly increased in the pharmaceutical industry, it is common for companies to conduct phase IV trials after a product has been launched. Such trials typically focus on further indications and subpopulations or seek to differentiate a product from its major competitors through head-to-head studies. Once valuable trial data have been generated, it is important to convey the information to key audiences, particularly opinion leaders, high prescribing physicians, or patients interest groups, in such a way that a product's benefits relative to its competitors become easily understood. In addition to a drug's profile, timing of market entry is another critical success factor.

1.4   Outlook

Some new markets are either not willing to pay the high prices common in the U.S. and most European countries, or they are highly uncertain, difficult and expensive to develop. One new market in the U.S. and European economies, for example, is geriatrics. People are becoming increasingly old, and healthcare spending is increasing with age. For example, the U.S. national health expenditures for the 65 years and older population was US$18,424 in 2010, five times higher than the spending for children (US$3628) and still three times higher than the average expenditure for the working-age population (US$6125). Thus, there are many new medicines under development that specifically aim at treating older people, targeting diseases such as diabetes, rheumatoid arthritis, Alzheimer's disease, or Parkinson's disease.

Focusing on blockbusters appears to be an enviable competitive position, given the strong first-mover advantages in the pharmaceutical market. However, if a significant share of total sales depends on blockbusters, a company exposes itself to the risk of a sharp sales decline once the underlying drug loses patent protection. Most pharmaceutical companies have thus started to balance and hedge their drug portfolios more rigorously and have diversified their businesses.

The Industry Challenge: Who Would Want to Be in This Business?

2

> *"We must never forget that innovation is an access issue—access for those with unmet medical needs. We must balance the needs of patients for marketed medicines today with the needs of patients depending on new medicines in the future."*
>
> Henry A. McKinnell, Jr.,
> Former President and CEO, Pfizer

2.1 A Highly Complex Industry

In good times, the pharmaceutical industry is a safe source of profits for its investors. But those good times are mostly a thing of the past, and several financial performance indicators have become markedly more pedestrian. The unique cocktail of competition from established and non-branded companies, regulatory organizations, new technologies, changing markets, and more recently also an increasingly wary public, makes business in the pharmaceutical industry difficult both for incumbent players as well as new entrant firms.

Of course, it is still possible to make good profits in this industry. But at the same time, product quality concerns and ensuing legal battles made pharmaceutical companies and drug approval administrations extremely cautious about launching new drugs. Large venerable firms were brought close to collapse in the wake of the Vioxx recall and other events. Emerging market companies in China and India start to compete with established pharmaceutical multinational firms not only on their home turf but globally, and the approval organizations in those countries become increasingly assertive and independent in their decision-making. What used to be an already highly complex industry has become even more complex. Maneuvering this industry is becoming increasingly difficult.

© Springer International Publishing AG, part of Springer Nature 2018
O. Gassmann et al., *Leading Pharmaceutical Innovation*,
https://doi.org/10.1007/978-3-319-66833-8_2

Industry Classification and Background

Historically, the pharmaceutical industry emerged from the chemical industry. Still today, chemistry represents a significant part in the innovation of pharmaceutical products. There are many similarities in the production processes of both chemical and pharmaceutical substances, and many industry experts aggregate the chemical and the pharmaceutical industries into one single industry called 'the pharmaceutical-chemical industry'. Despite the many similarities, there has been a general trend towards their separation into independent disciplines. Although there are still many companies that operate a pharmaceutical division alongside a chemical division, capital markets have increasingly favored firms that separated these businesses. The reasons are found in higher profitability and lower exposure to cyclic trends in the pharmaceutical industry, which leads to a stronger shareholder value orientation.

Today, pharmaceuticals are usually aggregated—along with the product groups of vitamins, fine chemicals, plant protection agents and animal medicine—under the broader category of so-called life science products (i.e., products that intervene in the metabolic processes of living organisms). Including specialty chemicals, the different product categories can be characterized as follows:

- The pharmaceuticals product group mainly includes patented, innovative products available only by prescription of a physician (either patented or generic). Over-the-counter (OTC) drugs and diagnostic aids have increased in importance as well.
- The product group of plant protection agents includes herbicides, fungicides and insecticides. The animal medicines group includes drugs for pets and livestock.
- The vitamins and fine chemicals product group includes the 13 vitamins and their derivatives as well as flavors and fragrances. These are not products for direct consumption, but rather 'bulk products' that are used for manufacturing pharmaceuticals, foodstuffs and animal feed.

The specialty chemicals product group comprises of several highly-specialized products that are frequently manufactured in relatively small quantities in response to specific needs of individual customers. Professional advice as a service to customers is of considerable importance.

The pharmaceuticals product group can be broken down further into several product groups that address various different therapeutic areas. The classification per therapeutic areas is often the basis for the organizational structure in most pharmaceutical companies.

Extensive Product Groups

Pharmaceutical products are defined as substances or mixtures of substances, which are meant for use in the recognition, prevention or treatment of diseases or for some other medical purposes regarding influences on the human organism (Leutenegger 1994). In general, drugs are differentiated into prescription drugs and non-prescription drugs. Further drug classifications include generic drugs, diagnostic drugs, orphan drugs or genetically manufactured drugs.

Prescription drugs are also often referred to as ethical drugs. They are only distributed by pharmacies or hospitals after they have been prescribed by a physician. In some cases, the physicians are also allowed to distribute prescription drugs themselves. These so-called self-dispensing physicians do not just prescribe the respective drug, they also dispense the product.

Non-prescription drugs can be purchased over-the-counter (OTC) at pharmacies and drugstores, or obtained in hospitals. Hence, non-prescription drugs include both medicines bought in pharmacies and drugstores without a prescription and medicines prescribed in medical practices and hospitals. OTC drugs are also sometimes referred to as drugs purely used for self-medication purposes (i.e., without any prescription at all). OTC drugs are usually used for minor ailments such as headache or the flu.

Generic drugs are chemically identical replications of prescription or non-prescription drugs whose patent protection has expired ("off patent"). Generic drugs (or "generics") are usually offered by firms that did not develop the drugs themselves but gained a license to sell the drug. As these firms do not need to recoup high R&D investments, generic drugs are usually marketed at a much lower price than the original drug, but they have the same efficacy because they use the same underlying substances as the original patent-protected drug. Drugs that are about to lose patent protection are thus exposed to intense competition. For instance, Eli Lilly's historic growth driver Prozac lost U.S. patent protection in August 2001, and generic competition almost immediately forced down sales by 66% in the fourth quarter of 2001. Many pharmaceutical companies have started their own generics business as a response to the generics challenge. Novartis acquired generic manufacturers Eon and Hexal and integrated them in its generics business operated under the brand name Sandoz, now one of the largest generics businesses in the world, and one of Novartis' global growth drivers.

Orphan drugs target rare medical conditions with usually very low patient populations. They provide physicians with therapeutic alternatives, and in some cases, they even provide a first therapeutic option. The FDA grants orphan drug status to a company if that drug is believed to substantially increase the life expectancy of patient in a given disease target. This excludes other companies from receiving an FDA license to produce a similar drug for a limited period (usually 7 years), thereby allowing the company producing the drug to regain their R&D expenses.

Orphan drugs do not generate large profits due to the small size of the target markets, and the developmental risks involved are still substantial. Hence, orphan drug legislation is perceived to generate advantages for the consumer. Still, in addition to exclusivity rights in particular indication areas, orphan drugs can help to improve the public profile of a company. In fact, the number of orphan drug applications continues to increase. The number of orphan drug designations grew from an average of 65 in the 1990s (when it was fairly constant) to 354 in 2015 (Karst 2017). Between 1995 and 2005, a total of 160 orphan drugs have been approved, and between 2001 and 2011, the market for orphan drugs grew at a CAGR of 25.8% (Gaze and Breen 2012). In 2014 the FDA approved the highest number of orphan drugs ever—49—and the approval numbers have been in the 30s and 40s since 2012 (Karst 2017).

Genetically manufactured drugs play an increasingly important role in the pharmaceutical pipeline. Gene technology includes all methods to characterize and utilize genetic material and is used for drug discovery, research simulation, and even diagnostic purposes. The limitations of gene technology are mostly ethical.

2.2   Five Forces Analysis of the Drug Industry

Given the strong dependence on innovation and growth, the uncertainty of product development success, and great complexity in pricing, marketing and distributing its products, it is perfectly reasonable to ask the question: How attractive is the pharmaceutical industry overall? This question cannot be adequately answered without addressing the balance of power among the various industry stakeholders. In addition, we must consider the exit barriers of existing competitors and the entry barriers of potential new companies.

Purely considering financial data, the pharmaceutical industry appears to provide comfortable returns, with a relatively high return-on-equity ratio (around 15% for pharmaceuticals, 22% for biotech in 2016) and one of the highest net profit margins (almost 18% for both pharma and biotech, according to Damodaran 2016). But one is misled to believe that the industry is characterized by little competition and safe and predictable environments.

A method that is often used to analyze industry attractiveness and identify opportunities and threats is Michael Porter's five-forces framework (Porter 1985; see Fig. 2.1). Porter summarized five principal forces that shape competition in an industry: the bargaining power of suppliers and buyers, the risk of entry from potential competitors, the threats of substitute products, and the degree of rivalry among established companies within an industry. A sixth, regulative force is often added to complement Porter's five forces in industries with a strong influence from regulation bodies; this is particularly true in the pharmaceutical industry and the related healthcare sector. Several opportunities and threats of the pharmaceutical players directly derive from public regulations, with direct impacts on pharmaceutical sales, margins, and costs.

Force 1: Bargaining Power of Suppliers

Suppliers in the pharmaceutical industry include providers of raw materials, biotechnology firms, and manufacturing plants, but also local co-marketing partners or the

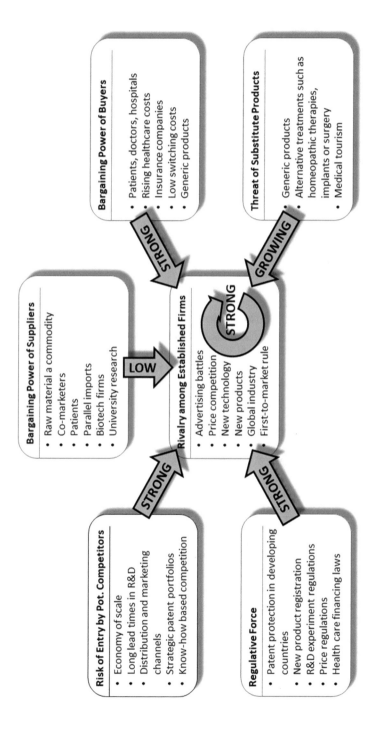

Fig. 2.1 The balance of power in the pharmaceutical industry

labor force. Pharmaceutical companies may have different suppliers depending on whether they compete in the OTC, ethical or generic businesses.

In the pharmaceutical industry, suppliers do not seem to have strong bargaining power. In clinical research, for instance, the suppliers are patients who participate in clinical trials, the investigators and their research teams who provide the data, and external contractors. An additional threat may emerge from parallel imports from low-price countries (which has triggered heated discussions over the prospects and dangers of adapting prices for life-saving but expensive treatments to developing economies and the possible threat of re-imports from these countries).

However, pharmaceutical companies are increasingly dependent on biotechnology companies and university research. The negotiation power of biotechnology companies has constantly grown over the last years, resulting in pharma-biotech deals that assign up to half of the profits and revenues to the smaller biotech partner today. However, pharmaceutical and biotechnology companies seem to prefer a co-existence rather than direct competition. While biotechnology companies provide early-stage input into the collaborations via new technologies or compounds, the pharmaceutical companies provide global marketing, broad and extensive distribution channels and a local sales force.

In addition, the biotechnology venture capital industry has matured in the past years which made it more difficult to acquire biotechnology start-ups (although some biotechnology companies have started to aggregate and announce their intention to become fully-fledged pharmaceutical companies, e.g., Celera and Axys Pharmaceuticals in 2002). While alliances have thus become an important source of new products and marketing agreements, the overall threat posed by biotechnology companies to pharmaceutical companies seems to be fairly moderate. It is generally not expected that biotechnology companies will turn into fully integrated pharmaceutical companies and compete via similar business models.

Force 2: Bargaining Power of Buyers

Buyers in the pharmaceutical industry usually include the patients (particularly in the OTC business), medical doctors who prescribe drugs, hospital boards who authorize the purchase of new treatments and drugs or pharmacists optimizing their stock of medication.

The ultimate consumer—the patient—usually does not have much influence on a medical doctor's or physician's decision about a certain prescriptive drug, since their knowledge about the respective drug and its consequences tends to be limited. Moreover, the patient normally does not carry the full costs of the product, at least not directly. These costs are typically covered by health insurance companies. Therefore, the consumer/patient has not developed a strong sense of cost consciousness regarding the pharmaceutical products being used. Hence, product quality is generally the main factor influencing the purchasing decision.

But this is likely to change with recent changes in national healthcare systems: Patients will participate in drug and therapy costs and, therefore, becomes more price sensitive. Co-payments for branded prescription drugs are a first step towards this direction. Subsequently, the informed patient is increasingly gaining in importance. In the case of non-prescription drugs (i.e., OTC drugs), the patients are usually able to select the drugs themselves. This leads to higher cost-consciousness on the consumer's side. Health insurance companies are thus increasingly opting for OTC drugs in order to reduce their own reimbursement obligation.

The emergence of self-organized patient groups requires pharmaceutical companies to pay more attention to direct-to-customer or direct-to-patient business models. While buyers of pharmaceutical products are still a highly-fragmented group, the buying power of large patient groups particularly in the U.S. has risen considerably over the last years. The so-called Health Maintenance Organizations (HMOs) aggregate and represent the interests of large patient pools. They have been very successful in negotiating prices that are favorable to patients. Thus, the buyers can exercise a strong influence over prices, by seeking price reductions for bulk purchases or threatening to switch to other suppliers (particularly in the generics business). The HMO model is particularly applicable in the U.S. where prices of prescription drugs are not subject to price regulation by the government.

Despite the threat of HMOs, most pharmaceutical companies have succeeded at establishing direct marketing relationships with doctors and patients around the world. Large groups of sales representatives and Medical Science Liaisons (MSLs) ensure appropriate awareness among physicians for new and existing products on the market. However, regulatory scrutiny has increased, and pharmaceutical companies carefully rethink and reassess their sales strategies. Particularly the high number of doctor visits by sales representatives has caused regulatory authorities to intervene. More focused approaches that deliver information with higher medical value to physicians are likely to represent the sales force model of the future.

The buying power of doctors seems to have increased slightly, and switching costs are relatively low. At the same time, governments and health authorities influence local prices in their attempts to contain healthcare costs. In countries with nationalized healthcare and tight price controls, buyer power is higher: Pre-scription prices in most European countries are about 30–50% lower than in the U.S. (Wall Street Journal 2015). According to BCG (2016), health care consumers have more money at stake, more choices to make, and more information—but also more noise!—to consider than ever before. Apart from primary physician referrals, word-of-mouth recommendation is as important as ever: In the U.S., 54% of all patients would ask doctors, 46% would ask friends and family before they book a minor procedure (BCG 2016).

Force 3: Risk of Entry from Potential Competitors

New entrants are usually faced with the following entry barriers:

- Economies of scale such as in R&D, marketing, and sales;
- Slow success rates in new drug development;

- Image, established relationships, and brand value;
- Capital requirements and financial resources;
- Access to distribution channels;
- Ability and capacity to deal with regulatory agencies and patents.

Even small biotech firms focusing on a single technology must spend hundreds of millions of dollars just to propose a potential product. Established pharmaceutical companies have manufacturing and distribution systems that are hard to replicate, a strategic patent portfolio preventing competitors to enter new disease areas, and large marketing budgets to protect their brands. Nevertheless, some companies are trying to enter the pharmaceutical market, such as contract research organizations who attempt to expand their businesses vertically, i.e. downstream into the pharmaceutical industry.

Furthermore, patent protection does not protect against competition from generic products. While generics capture about 50% of unit sales with continuous growth, they are far less profitable than ethical drugs. For instance, Pfizer's 2001 prescription drug revenues of US$26.3 billion was almost five times the combined sales of the 11 leading generic drug makers covered in a Standard & Poor's industry survey (Saftlas 2001).

Force 4: Threat of Substitute Products

Substitute products perform the same function as existing products, and sometimes even better. Generics are substitutes for the original drugs but at a lower price. While generics mount an increasing threat to profitability of large pharmaceutical companies, they also offer opportunities. Novartis, for example, has proactively approached the threat of generics, and has emerged as one of the largest generics companies itself by selling various generic products under the global umbrella name Sandoz.

The worldwide generics market has experienced significant growth in unit sales. However, the market share of generics is still comparatively small compared to the total drug market due to the generic drugs' relatively low prices. A reason might be that manufacturers of generics typically concentrate on drugs with very high sales volumes to reach critical mass in sales and margins quickly, thereby ignoring smaller niche markets that have been highly profitable for pharmaceutical companies.

Besides generics, other substitutes for pharmaceutical products include certain medical devices or alternative therapies. Even hospitalization may be a substitute for drug treatments. For instance, surgery may make drug intervention unnecessary. Of course, these are not perfect substitutes, and many factors need to be considered in the medical decision making. On a cost-to-value basis, however, surgery, prolonged medical care and hospitalization are less attractive.

In many countries, new substitute products will be financed by the insurance firms only if the added value of the new product is proven to be much higher than the old product. Using such regulations insurance firms try to reduce costs as new products are typically more expensive than older ones.

Alternative therapies such as homeopathic remedies, acupuncture, and herbal medicines are all still considered medically unproven and are usually not covered by health insurance. But traditional treatment knowledge might change with a new generation of medical doctors who are educated more openly and are trained to consider certain patients' wishes for soft treatments. Should these products demonstrate medical efficacy, they will be quickly absorbed into conventional medicine. Overall, the risk of unmanageable exposure to substitute products in the pharmaceutical industry is relatively low. The strongest product substitutes still come from the innovative pharmaceutical companies themselves.

However, in the next two decades serious competition is likely to arise from current industry outsiders. For instance, the consumer electronics industry (e.g., Apple) are poised to attack the preventive healthcare sector. Life style companies presently outside the attention of the FDA are going to become a source of substitute products in the near future. In addition, the trade with mainly unstructured 'big data' is already attracting information-based companies (e.g., Google) and insurance companies. Serious competition to the drug as-we-know-it might arise from the digital pill. Universities such like MIT, Dartmouth, Stanford and St. Gallen/Zurich are already expanding their research programs into clinical trials on digital diagnostic and therapies in areas such as obesity, Alzheimer's, and depression. Digital therapies are attractive for therapies with high volume and longitudinal character, i.e., areas in which machine learning and artificial intelligence work best.

Force 5: Rivalry Among Established Companies

The rivalry among incumbent pharmaceutical firms is moderately intense since the pharmaceutical industry is still quite fragmented. The top-ten drug manufacturers controlled 42% of the global market in 2000, and 46% in 2007, but their share has since dropped to 35% in 2014. Competitors have eroded their stronghold by means of price competition, acquisitions, advertising battles and new product introductions. This rivalry is particularly intense in saturated markets (e.g., cardiovascular and central nervous systems, pain relievers), and less intense in growing markets (e.g., oncology or immune disorders).

Most industry profits come from patented products or therapies. The so called 'me-too' products tend to be less profitable. As individual drug therapies are generally quite focused on specific markets, competition is somewhat limited. Nevertheless, generics are improving their position at the expense of blockbuster drugs going off-patent (experiencing price drops of up to 80%).

Rivalry is typically fought over time-to-market, since first-to-market companies gain a relatively high market share and thus are more likely to recoup R&D and marketing expenses. Global product introductions help achieve market share, and therefore large companies have an edge over smaller companies thanks to more developed marketing and distribution systems.

Force 6: The Regulators

Public laws and regulations play perhaps a greater role in the pharmaceutical industry than in any other. The regulative force impacts pharmaceutical innovation on several levels:

1. R&D regulations and product registrations,
2. Price regulations and national healthcare systems, and
3. Intellectual property rights.

R&D Regulations

R&D on drugs, from research to experiments and clinical trials, are to be conducted in compliance with regulations issued by relevant agencies such as the **Food and Drug Administration** (FDA) in the U.S. or the **European Medicines Evaluation Agency** (EMEA) in Europe. These governmental agencies stipulate authorization and registration procedures for all new drugs submitted for approval in their respective markets. New drugs must prove that they are suitable for use in human beings and the respective benefit-risk profile must be determined prior to marketing approval. Only after a medical product has cleared all hurdles—and therefore fulfills regulations regarding quality, efficacy, and safety—is it granted authorization.

The FDA, for example, interferes at two stages during a new drug authorization/registration process. The first comes right after the pre-clinical tests, where the FDA's Center for Drug Evaluation and Research (CDER) determines if the new drug is suitable for use in clinical trials (i.e. in trials with humans). This process is called the Investigational New Drug (IND) Review Process. An IND permission must be maintained annually by sending, for example, annual reports. In addition, projects are discussed with the FDA frequently, especially before entry into phase III clinical trials and submission. After the successful conclusion of clinical tests, and prior to approval for marketing, the CDER determines the benefit-risk profile of a new drug. This process is called the New Drug Application (NDA) Review Process.

In general, examining pharmaceuticals for safety, quality and other pre-market considerations is time consuming and expensive. However, the FDA uses timesaving processes to speed up introduction of important new drugs to patients with particular needs. An accelerated approval may be granted to priority drugs that show promise in the treatment of serious and life-threatening diseases for which there is no adequate therapy. Treatment IND designations enable patients not enrolled in the clinical trials to use promising life-saving drugs while they are still in the testing stage. For example, when the first tests of the antiviral drug AZT showed encouraging results in 330 AIDS patients, the FDA authorized a Treatment IND for more than 4000 people with AIDS before AZT was approved for marketing.

The median length of time required to review and approve a new drug varies by country, from 1.1 years in the U.K. up to 1.7 years in Germany, Australia, Spain, and the U.S. Legal regulations often reflect the entire society's position towards technology. In highly-developed industrial countries, public acceptance of new technologies, such as bio- and gene-technology, is in decline, with restrictive regulations for experiments on animals and stem cell research as typical examples. Thus, pharmaceutical companies consider regulations as a driver to shift some of their research abroad. Besides approvals for clinical trials in humans, animal trials and inventions in gene technology are covered by strict authorization processes as well.

Price Regulations and National Healthcare Systems

In most countries, drug prices are regulated by federal authorities (directly or indirectly). The United States and New Zealand are the only two countries in the world that have no federal price regulation for drugs. In some countries, the price of a product is fixed relative to the social costs of its people. Yet in other countries, the price of a drug is defined by its innovativeness as measured by the number of patents in that area (e.g., in Brazil). However, national healthcare systems always have the primary and most direct impact on product prices, especially when reimbursed by health insurance organizations.

This process is perhaps best observed, and most analyzed, in the United States, where the Trump administration has made it its goal to reverse the Obamacare initiatives made in the previous legislative period. Several questions are particularly critical for pharma companies: Will the Trump administration support further innovation in payment models? Will M&A regulations intensify? The BCG (2016) report speculates that regulatory oversight of medtech and biopharma is likely to expand. Politics and pharma have always been very interconnected, and the impact on business models cannot be overstated.

Intellectual Property Rights

The purpose of patent law is to support the development and protection of new technology and know-how. On the one hand, inventions need to be made available to the general public. On the other hand, innovators must be reassured that their inventions are protected against unlawful imitation and replication. Pharmaceutical patents are essential because it is not difficult to ascertain the respective substances of a drug and, consequently copy or imitate pharmaceutical products. Studies have shown that patents are the most effective means of appropriation. 65% of pharmaceutical inventions would not have been introduced without patent protection, compared to a cross-industry average of 8% (Reuters 2002).

Patent protection is unclear or still under development in some key areas of pharmaceutical R&D. For instance, there are differences to what extent genes can be patented (and thus 'owned') in the U.S., Europe, and Japan. In the U.S., naturally occurring genes are no longer patentable since 2013, but synthetic DNA (or complementary DNA, or cDNA) can be patented. This is important as cDNA is crucial for producing man protein-based drugs. Sometimes, international patent law is only accepted if national interests are maintained. Brazil, for example, has threatened several times to suspend domestic compliance with international patent rights for malaria drugs unless certain license fees were dropped.

2.3   Pressure on the Traditional Pharma Business Model

Pharmaceutical companies not only need to outpace others through innovation and R&D, their business models are also challenged more fundamentally. The traditional model of spending substantial amounts on developing a new drug, and then enjoying quasi-monopoly pricing for the remainder of the drug's patent life, is coming under serious pressure. This pressure is driven by:

* Continued rise of generics/branded generics products
* Relevance of specialty pharma drugs
* Pressure for portfolio optimization
* Increasingly competitive marketing efforts
* Legal challenges

Continued Rise of Generics/Branded Generics Products

Generics drugs, i.e. low-cost copies of branded drugs that usually have lost patent protection, are nothing new to the pharmaceutical industry. Priced at significant discounts of 50–70% to their branded counterparts, generic drugs typically capture significant market share within months after their introduction, and thus pose significant risks to established pharma companies' product portfolios.

Due to their competitive pricing, several health plans and governments around the world actively encourage the use of generic drugs in the effort to curb ever-increasing healthcare costs. According to IMS research (IMS 2016), generic drugs account for 88% of all prescriptions filled in the U.S. in 2016, and they may account for over 90% of prescription volumes by 2020. Generic drugs are expected to account for 52% of global pharmaceutical spending growth from 2013 to 2018, compared to 35% for branded drugs. Overall, sales of generic drugs are forecast to reach $442 billion in 2017, an annualized growth rate of 10.6% from 2013 to 2017.

Generic drug companies operate very differently from traditional pharma companies. Their investment in R&D is significantly lower, because the drug has already been developed and approved. Several companies, such as Teva or Mylan, have emerged as significant and highly successful players in the generics space, and they continue to capture value from traditional pharmaceutical companies. While generics companies typically introduce their generic products only once a branded drug loses its patent protection, some generics companies now introduce their generic products even before a comparable branded drug loses its patent protection. Legal challenges are hereby proactively taken into consideration.

One fairly successful strategy pursued by generics companies is referred to as selling 'branded generics'. The branded generics strategy is popular as it achieves the best of both worlds: keeping R&D costs low by selling generic products, while at the same time achieving a sizable margin on these low-cost products by positioning them as branded drugs. With this strategy, generics companies do not only capture market share from traditional drug companies, they also try to maximize value extraction from the market.

Relevance of Specialty Pharma Drugs

Therapeutic areas are becoming increasingly complex. As a result, the corresponding pharmaceutical drugs are becoming costlier, because they oftentimes treat complicated (e.g., cancer, rheumatoid arthritis) or rare chronic conditions (e.g., HIV). The corresponding drugs for these complex conditions are usually referred to as specialty drugs.

Specialty drugs are typically higher-priced than traditional drugs due to their smaller patient populations and higher complexity. For instance, Medicare defines a specialty drug any drug for which the negotiated price is US$600 per month or more. Drugs are also identified as specialty when there is a special handling requirement or the drug is only available via a limited distribution network. The larger price premiums of specialty drugs are typically explained by the perceived value of rare disease treatments which usually is very expensive when compared to treatments for more common diseases.

In 1990, there were 10 specialty drugs on the market, in the mid-1990s there were fewer than 30, and by 2015 there were 300 specialty drugs on the market (Kober 2008). According to Thomas and Pollack (2015), specialty medications accounted for one-third of all spending on drugs in the United States in 2015, up from 19% in 2004 and heading towards 50% over the next 10 years. Unsurprisingly, the success of specialty drugs comes at a time when the traditional pharma business of selling blockbuster drugs to mass markets has started to experience problems.

In many cases, specialty drug therapy areas are served by smaller and highly specialized companies. Often the players in these markets aim to occupy niches by offering a range of treatment options in their respective specialty areas. Sometimes the specialty pharma companies also acquire active ingredients from adjacent therapeutic areas and apply them under different indications in their respective niche markets. Well-known specialty pharma companies include Allergan or Shire.

One company that is well known for a successful specialty pharma strategy is **Valeant Pharmaceuticals**. Valeant aggressively acquired various substances and then marketed them in several high-priced niche markets. By pursuing this strategy, Valeant was able to keep R&D costs at a minimum and thus created significant value for its shareholders. This strategy worked well for several years, but in August 2015 questionable aspects were revealed about Valeant's business practices, including the company's ties to the 'captive pharmacy' Philidor. Subsequently, Valeant experienced a significant drop in its share price including a senior management shakeup that also partially eroded confidence in the specialty pharma business model.

Nevertheless, the specialty pharma business model is likely to have promising prospects for the future, and is expected to have significant impact on the structure and approach how many pharma companies will innovate.

Pressure on Portfolio Optimization

Traditionally, big pharma companies have focused on large therapeutic areas in order to increase their likelihood of finding blockbuster products. Over the last several years, however, many pharma companies have started to restructure their drug portfolios towards focused therapeutic areas, in the hope to have a higher chance of offering first-in-class medications or where they could meet large unmet medical needs.

One such area is oncology, and Fig. 2.2 shows why so many pharma companies are currently focusing on it as their primary target: With US$83 billion in pharma sales in 2015 and expected to grow at 12.5% p.a. to US$190 billion by 2022, oncology represents by far the most attractive therapeutic market opportunity. Many pharma companies have recognized this trend early and positioned themselves as market leaders in this therapy area.

For instance, as of 2016, **Roche's** clinical pipeline was focused on only five major therapy areas, including inflammation/immunology, neuroscience, infectious diseases, ophthalmology, in addition to oncology. Also in 2016, **Novartis** announced to split its pharmaceuticals division into two business units reporting to the CEO: Novartis Pharmaceuticals and Novartis Oncology. According to Novartis' press release, the new structure reflects the importance of oncology to Novartis following the successful integration of the oncology assets acquired from GSK. Novartis expects this change to help drive its growth and innovation strategy. Many other companies in the industry have also streamlined and focused their portfolios in order to maximize value extraction. These strategic portfolio considerations need to be taken into account when defining a company's innovation strategy.

Increasingly Competitive Marketing Efforts

While innovation and R&D budgets are without any doubt one of the most important drivers for value creation in the pharmaceutical industry, it should not be forgotten that marketing is another very critical spending category. Marketing budgets often exceed R&D budgets, with nine out of the ten largest pharma companies spending more on marketing than on innovation (see Fig. 2.3).

This spending behavior and the resulting interaction between R&D and marketing budgets has several implications that need to be taken into account when talking about pharmaceutical innovation: When considering the imperative to launch blockbuster products, it becomes evident that associated "blockbuster costs" are not sustainable without corresponding "blockbuster revenues". Therefore, increasingly competitive innovation activities require increasingly competitive marketing efforts. Considering actual R&D and marketing costs, it can be argued that today's R&D decision-making is largely driven by marketing considerations.

But the time of huge marketing spending for individual drugs may be coming to an end, not least because of the trends discussed above: more focused portfolio approaches, the relevance of generics products, and the rise in niche and specialty

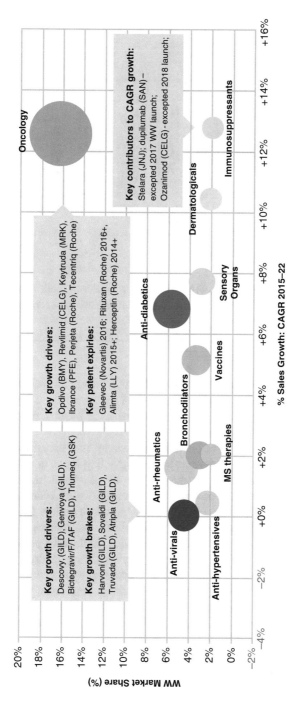

Fig. 2.2 Top ten therapy areas in 2022 by market share and sales growth. Source: Evaluate (2016)

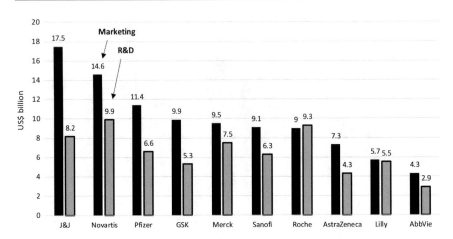

Fig. 2.3 Marketing and R&D spending at major pharma companies in comparison. Source: Global Data, 2013 values

pharma applications targeting smaller patient populations. Highly targeted marketing strategies aligned with specific patient populations are expected to take over.

Legal Challenges

The ever-increasing pressure for growth and new product launches in the pharmaceutical industry has, unfortunately, led to several concerns from a legal standpoint. Several cases of incorrect and inappropriate behavior by market participants in recent years are raising concerns.

One of the biggest issues in the pharmaceutical industry is the off-label promotion of drugs. Off-label marketing concerns the selling a drug for an unapproved indication or in an unapproved age group, dosage, or route of administration.

Over the last several years, pharma companies have spent billions of dollars in fines relating to off-label product sales. In the biggest such settlement in U.S. history so far, **GSK** paid US$3 billion to resolve criminal and civil charges related to illegal promotion of drugs and failure to report safety data. At the core of the controversy was the antidepressant Paxil. GSK pleaded guilty for selling the drug as a treatment for children under 18, even though the FDA had never approved the drug for this use. According to a 2016 Public Citizen study, federal and state governments and pharmaceutical manufacturers reached a total of 373 settlements totaling US$35.7 billion in the U.S. between 1991 and 2015. Table 2.1 lists some of the biggest pharma settlements in marketing fraud history.

Drug Watch estimates that off-label drug sales contribute to about US$40 billion (or about 20%) in revenues each year. In the case of Risperdal, for instance, off-label prescriptions were argued to account for 75% of sales in 2002. The U.S. Department of Justice reached a settlement of this case only in 2013, more than a decade later.

Table 2.1 Biggest settlements in marketing fraud in the pharmaceutical industry

Company	US$ Settlement	Drugs	Year
GlaxoSmithKline	3.0 billion	Paxil, Wellbutrin, Avandia	2012
Pfizer	2.3 billion	Bextra, Geodon, Zyvox, Lyrica	2009
John & Johnson	2.2 billion	Risperdal, Invega, Natrecor	2013
Abbott	1.6 billion	Depakote	2012
Zyprexa	1.4 billion	Zyprexa	2009
Merck	950 million	Vioxx	2011
Serono	704 million	Serostim	2005
Purdue Pharma	634.5 million	OxyContin	2007
Allergan	600 million	Botox	2010
AstraZeneca	520 million	Seroquel	2010
Bristol-Myers Sq.	515 million	Abilify	2007

Source: Drug Watch

In addition to off-label marketing, other forms of inappropriate market behavior have shaken up the pharma industry. One prominent example is **Turing Pharmaceuticals**, which made negative headlines in 2015 when its CEO Martin Shkreli raised the price of its drug Daraprim from US$13.50 to US$750 per pill overnight, an increase of approximately 5500%. Many patients—many of them elderly retirees—were no longer able to afford this medication. This price gouging not only made the CEO the "most hated man in America," it also triggered investigations by the U.S. Congress into drug pricing practices and increased overall scrutiny by government officials.

2.4 Growth Drivers for the Years to Come

We are the victims of our own success. People are getting older, which brings about afflictions associated with age, changing lifestyle, and opportunity for diseases to occur in the first place. This is even more evident in the so-called emerging markets in Asia. These trends drive up healthcare costs, and pharmaceutical companies will benefit from these trends as they will use new breakthrough therapies, diversification, access to new regional markets and M&As to expand.

The Economist Intelligence Unit (EIU) predicts that the global healthcare market will grow at 5.3% per year between 2014 and 2018 (Nicholls and Brayshaw 2014). North America will grow by 4.9% annually, Western Europe at a rate of 2.5%, Latin America on average by 5.0%, Asia by 8.3%, and Middle East and Africa by 8.7%. The increase in life expectancy generates demand for drugs to treat age-related disease, such as Alzheimer's disease or arthritis. Lifestyle-related disease, such as diabetes, are also on the rise, as people in the in developing economies, such as in Mexico, India and China, adopt Western life styles, change diets and become obese.

Global pharmaceutical sales are forecasted to increase on average by 6.7% per year (2014–2018) and achieve US$1.61 trillion in 2018 (Nicholls and Brayshaw

2014). Key drivers of this growth are the Obama healthcare reform, rising prices for innovative drugs and the dynamics of the Asian market, such as in China, India, Indonesia, Malaysia, South Korea and Thailand. For example, China became the second largest pharmaceutical market worldwide in 2016. Worldwide prescription drugs sales are estimated to increase at around 5% (CAGR 4.8%) from US$743 billion in 2014 to US$987 billion in 2020 (see Fig. 1.9). The worldwide sales of generics alone will grow from US$74 billion in 2014 to US$112 billion in 2020, increasing its market share from 7.4% in 2014 by revenue of the global pharma market to 11.3% in 2020 (Evaluate 2015). The number of high income households (household with an annual income of more than US$25,000) is forecasted to increase to around 570 million, indicating a record number of people with disposable incomes high enough to afford innovative drugs (Deloitte 2014).

Innovation will continue to be one of the major growth factors in the coming years. The number of NMEs approved by the FDA increased remarkably in the past 2 years compared to the decade before (see Fig. 1.6). Breakthrough therapies have been a major driver of growth in the industry in the last years, exemplified by Abbvie's Humira or Gilead's Solvadis. There are indications that this trend will continue:

- Worldwide pharmaceutical and biotechnology R&D spendings are forecasted to increase from US$142 billion in 2014 to US$160 billion by 2020 (CAGR: +2%, see Fig. 1.1).
- The industry's R&D pipeline is valued at around US$500 billion, illustrating a positive perspective for the future (Evaluate 2015).
- According to a study by PwC, there is a clear correlation between innovation and growth, as the most innovative pharmaceutical companies (top 20%) included in the study grew at a rate of 16% faster than the least innovative (Arlington and Davies 2014). In the same study, the interviewed executives of major pharmaceutical companies mentioned that their companies will focus on product innovation rather than service or business model innovation, and 45% of the interviewees expected that they will provide further breakthrough drugs to the market. For example, Vertex's Orkambi (for the treatment of cystic fibrosis), Novartis' LCZ696 (to treat heart insufficiency), or Pfizer's breast cancer Ibrance are expected to become blockbuster drugs and to be in the top 50 of best-selling products by 2020.

The global sales contribution from biologics is predicted to increase from 23% (2014) to 27% (2020) (Evaluate 2015). Biologics are expected to revolutionize the treatment of autoimmune disease and cancer, two market segments that already grow rapidly, providing additional growth in the pharma sector. Biologics represented "only" 21% of the top 100 drugs by sales in 2006, but this share is forecast to represent 46% by 2020.

Although some commentators of the industry assert that the blockbuster model is broken, recent predictions suggest that worldwide prescription drug sales of the top 50 drugs will increase from US$161 billion in 2014 to US$233 billion in 2020 at a

CAGR of 6%. This translates into a significant increase of the financial value of blockbusters and their importance for the future of the industry. Eight of the top ten drugs that have been approved in 2014 are forecast to become blockbusters within 5 years of their launch (Evaluate 2015).

In general, cancer drugs have the potential to become blockbusters because of the high medical need to treat this life-threatening disease. Oncology will be by far the leading therapeutic area by sales in the pharmaceutical sector by 2020. As illustrated in Table 1.6, it is forecast that cancer drugs will account for US$153.1 billion in 2020, which is a CAGR of 11.6% compared to 2014. Although blockbuster drugs such as Gleevec, Rituxan or Herceptin already lost or will soon lose their market exclusivity, new billion dollar drugs are already in the pipeline for launch and bolster the growing market potential of oncology. The other main therapeutic areas are anti-diabetics (US$60.5 billion, CAGR +6.4%), anti-rheumatics (US$53.2 billion, CAGR +1.5%) and anti-viral drugs (US$49.6 billion, CAGR +2.3%).

In 2020, each of the top 50 drugs is forecast to provide worldwide product sales of more than US$2.5 billion. According to Evaluate (2015), the top selling drug is expected to be Humira (US$13.934 billion, by Abbvie) followed by Revlimid (US$9.640 billion, Celgene) and Opdivo (US$8182 billion, Bristol-Myers Squibb). Measured by forecast global prescription drug sales in 2020, Novartis (US$53.3 billion), Pfizer (US$44.9 billion) and Roche (44.7 billion) will retain their stronghold in the sector, as will pharmaceutical companies from the U.S. in general with five of the top ten companies headquartered there (see Fig. 2.4).

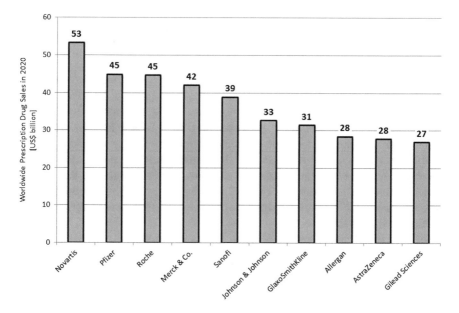

Fig. 2.4   Top ten pharmaceutical companies by worldwide prescription drug sales in 2020. Source: Evaluate (2015)

2.5 Seeking Salvation in M&A and Diversification

After the industry saw these and other major deals taking place in the 2000s, many industry observers thought that the time of big M&As were over. However, M&A activity continued to be strong, with potential mega mergers such as Pfizer's attempt to acquire AstraZeneca in 2014, Abbvie's attempt to take over Shire in 2014, or Pfizer's failed deal with Allergan in 2015, and completed acquisitions of e.g. Wyeth by Pfizer in 2009, or Roche merging with Genentech in 2009.

After 456 global deals in 2012 and 615 announced M&A deals in 2013, the industry saw an increase in deal numbers and sizes also in the last 2 years with a total volume of M&A transactions globally of US$300 billion. Recently, Chinese multinationals started looking for technology acquisitions across different industries (e.g., ChemChina's recent acquisition of Syngenta, a Swiss agrochemicals company in 2016/2017). In addition to the multi-billion dollar mergers, pharmaceutical companies are also acquiring biotechnology companies to access pipeline projects, technologies, IP rights and know-how.

The biggest M&A transactions in the industry so far have been:

- Pfizer acquired Warner-Lambert in 1999 valued at US$87.3 billion,
- Sanofi acquired Aventis SA for US$73.5 billion in 2004,
- Glaxo acquired SmithKline Beecham for US$72.4 billion in 2000,
- Allergan acquired Actavis for US$65 billion in 2015,
- Pfizer acquired Pharmacia for US$64.3 billion in 2002, and
- Pfizer acquired Wyeth for US$64.2 billion in 2009.

Case in Point: Mergermania Around Actavis/Allergan

Actavis (formerly Watson Pharmaceuticals) is a global pharmaceutical company that grew significantly through mergers and acquisitions (see Fig. 2.5). For instance, Actavis acquired the specialty pharma company Forest Laboratories for US$23 billion and the Botox company Allergan for US$66 billion in 2014. Although Actavis was headquartered in Dublin, Ireland before the merger with Allergan, the deal between Actavis and Allergan (a U.S. company) has not been ranked as a tax inversion deal. The new firm's sales are supported by the blockbuster drug Botox, which generated global sales of around US$2 billion in 2013. The resulting new **Allergan** has worldwide revenues of US$15 billion and thus one of largest pharmaceutical companies by sales.

In 2015 **Pfizer** announced plans to acquire Allergan (after Pfizer's failed attempt to acquire AstraZeneca the year before). The US$160 billion deal would have been the third biggest M&A deal made ever across all industries. Pfizer would have become the world leading pharmaceutical company with expected total revenues of US$74 billion after Novartis took over the industry leader role in 2014. Financial considerations seem to have been crucial in these acquisition plans. It was argued that the deal would have helped Pfizer to lower tax payments, as Allergan's

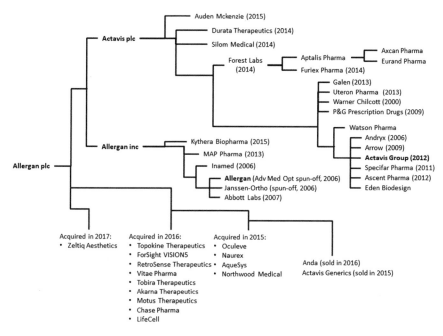

Fig. 2.5 Allergan's recent M&A history spans dozens of companies. Source: Allergan website

headquarter was located in Dublin in Ireland, a country with low corporate taxes. In April 2016, however, Pfizer terminated the deal with Allergan, after the U.S. Treasury Department raised concerns about U.S. companies to move overseas purely for tax reasons. Under the terms of the termination agreement, Pfizer agreed to pay Allergan US$150 million for expenses associated with the failed deal.

Motivations for Pharma M&A

One of the most frequently offered reasons for mergers is the exploitation of synergistic effects, resulting in the reduction of costs for infrastructure, administration, sales, taxes or even R&D. Access to new regional markets and industry subsectors are also important. Other key drivers for M&A deals are:

- Compensation of revenues losses by blockbuster patent expirations,
- The explosion of technology-based treatment innovations and core competencies,
- The need to fill R&D pipeline gaps,
- The aim to access strategically important IP, and
- The need to seek for growth options such as the access to new regional markets and the strengthening of the distribution channels.

Pharmaceutical companies pursue a mix of defensive and aggressive growth strategies. Defensive strategies aim to retain a competitive position by means of co-marketing agreements, co-selling, cross-licensing, and market-related acquisitions. This strategy builds critical size and momentum, reduces costs of sales, and essentially creates entry barriers for newcomers. Aggressive strategies try to overcome market entry barriers set up by competitors. While this strategy does not immediately reduce overall costs, it does provide a greater potential for creating value-added and hence long-term cost reduction. M&A transactions can support both kind of strategies when compensating revenues losses in case of patent expirations, filling the pipeline when R&D gaps arise, accessing strategically important IP when set as a market barrier by a competitor, accessing new regional markets when other growth options fail, or even saving taxes to increase profit and strengthen the financial capabilities of the company.

Thus, it becomes apparent that mega M&A deals will still play an important role in the growth strategies of some major research-based pharmaceutical companies. It will be fascinating to see how Pfizer will achieve its growth objectives after the failed mergers with AstraZeneca and Allergan, as it will be interesting to follow-up the developments of the Swiss pharma giants Novartis and Roche. Both companies are headquartered in Basel and Novartis is one of the main shareholders of Roche with an equity stake of 33.3% of Roche.

Growth Through Diversification

Diversification is another option to increase revenue streams. For example, Eli Lilly acquired the animal health division of Novartis for US$5.4 billion in 2015, creating the second largest animal health company and adding around US$2 billion to Eli Lilly's annual total sales. Pfizer acquired a lucrative vaccines business when merging with Wyeth in 2009, enlarging its product and R&D portfolios with this forward-looking technology. Today, Pfizer business comprises branded and generic prescription pharmaceuticals, over-the-counter (OTC) medicine and vaccines. Bayer started to increase diversification of its pharmaceutical business already in 2004 when acquiring the OTC business of Roche. Since then, further acquisitions have strengthened and broadened Bayer's portfolio, such as the US$1.2 billion acquisition of the U.S. vitamin and supplement maker Schiff Nutrition International in 2015.

A smart diversification strategy can pay off. For example, **J&J** generates 40% of its revenues from pharmaceutical sales, 40% from medical devices and 20% from consumer products mitigating the risks associated with the different business areas, providing J&J excellent growth options and group results. J&J was the most valuable pharmaceutical company (market value of US$277 billion) and recorded the highest net income (US$17.1 billion) amongst all pharmaceutical companies in 2014 (Evaluate 2015).

Given the trends towards an ageing and a generally more health-concerned society, pharmaceutical companies will keep their leading role in the foreseeable future. The top 20 pharma companies will spend almost US$100 billion in 2020, and the top 8 more than US$5 billion each in R&D (Evaluate 2015).

2.6   Conclusions

The pharmaceutical industry covers a broad range of product groups, from vitamins, fine chemicals, plant protection agents and animal medicine to the traditional pharmaceuticals and biopharmaceuticals. Within the group of pharmaceuticals, there are several different product categories that address multiple different therapeutic areas.

But the high complexity of the industry itself does not seem to be biggest hurdle for company success. Despite potentially attractive profit margins and a favorable balance of power with respect to suppliers and product substitutes, the pharmaceutical industry is a tough industry to be in overall. What used to be a 'safe haven' for market leaders has developed into a substitution market. While many diseases still await effective treatment, most of the obvious or easy drug targets have been discovered, and economically viable therapies have been developed. Start-ups and other entrants employ new technologies to compete with incumbent pharmaceuticals over international markets and new innovations.

While patent protection usually secures a monopolistic market power for a limited time, the threat of generics is immense. Sales of branded products that took years to be established collapse within weeks. Most pharmaceutical companies still do not know how to handle this competitive threat.

The Science and Technology Challenge: How to Find New Drugs

3

> *"The stakes are too high, the medical need too great, and the science too compelling for all of us to do business as usual."*
>
> Mikael Dolston,
> President of Woldwide R&D, Pfizer

3.1 Rise of the Biotechnology Industry: Boosting Innovation

Due to the enormous costs of building up a pharmaceutical R&D infrastructure, it was generally believed until the 1980s that no new company would ever be able to enter the pharmaceutical business and compete with the established industry's giants (see Robbins-Roth 2001). Ironically, this is when entrepreneurs created an entirely new industry—the biotechnology industry. In addition to an innovative management approach and creative funding strategies, the main drivers for the rise of the biotechnology industry was the emergence of new sciences and technologies. Established pharmaceutical companies were traditionally based on medicinal chemistry, physicochemistry and pharmacology, but biotechnology companies built their reputation in many novel areas, such as cell biology, molecular genetics, protein chemistry and immunology (Whittaker and Bower 1994).

The application of biotechnology in the pharmaceutical industry started with the development of scientific techniques such as genetic engineering and antibody production. Genetic engineering was developed in 1973, and it received its first commercial pharmaceutical application 4 years later when Eli Lilly started the development of recombinant human insulin in cooperation with Genentech. The resulting product, Humulin, became the first biotechnology product when launched in 1983. Over the past years, the number of biologics increased and today around nearly 50% of all FDA-approved new drugs are biologics (see Fig. 3.1). By the end of 2013, a total of 91 biologics received approval by the U.S. American Food and

© Springer International Publishing AG, part of Springer Nature 2018
O. Gassmann et al., *Leading Pharmaceutical Innovation*,
https://doi.org/10.1007/978-3-319-66833-8_3

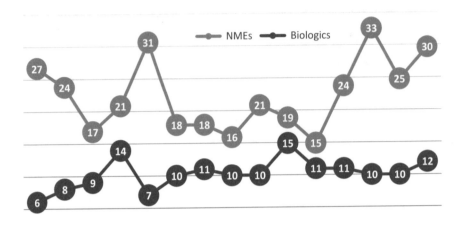

2000 2001 2002 2003 2004 2005 2006 2007 2008 2009 2010 2011 2012 2013 2014

NME new molecular entity
Source: Evaluate (2015)

Fig. 3.1 NMEs and biologics approved by the FDA (2000–2014)

Drug Administration (FDA), of those 34 were monoclonal antibodies, 26 enzyme modulators and 31 receptor modulators (Kinch 2015).

The proportion of biologics as part of worldwide prescription drug sales increased respectively. While biologics accounted for only 14% of all sales globally (US$78 billion) in 2006, they already accounted for 24% in 2015 (US$183 billion), and it is expected that 27% of the worldwide prescription drug and OTC pharmaceutical sales will come from biologics by 2020 (US$278 billion) (see Fig. 3.2).

The first biotechnology drugs that reached blockbuster status and generated in excess of US$1 billion in sales per year were Procrit (J&J), Epogen (Amgen), Neupogen (Amgen), and Humulin (Genentech and Eli Lilly). Today, seven out of the top ten best sellers worldwide are biologics (see Table 3.1). Humira (adalimumab), an anti-tumor necrosis factor alpha monoclonal antibody, was the best-selling product globally with total worldwide sales of US$12.890 billion in 2014.

Accordingly, and as illustrated in Table 3.2, the companies with the highest worldwide prescription drug sales derived from biotechnology products in 2014 were Roche (US$30.8 billion), Amgen (US$17.6 billion), Sanofi (US$16.0 billion), Novo Nordisk (US$15.0 billion), AbbVie (US$13.4 billion), Pfizer (US$10.8 billion), J&J (US$10.6 billion), Merck & Co. (US$8.2 billion), Eli Lilly (US$6.0 billion) and BMS (US$3.7 billion). Evaluate Pharma (2015) forecasts that these companies will further increase their proportion of biotechnology drug sales at compound annual growth rates (CAGR) of 2–23% per company between 2014 and 2020.

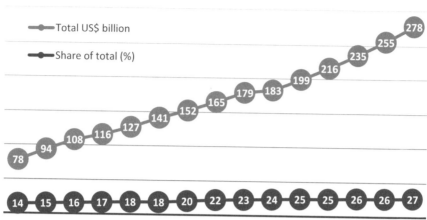

2000 2001 2002 2003 2004 2005 2006 2007 2008 2009 2010 2011 2012 2013 2014
Rx prescription, *OTC* over-the-counter
Source: Evaluate (2015)

Fig. 3.2 Worldwide prescription (Rx) and OTC sales of biotechnology products

The advancement of biotechnology impacts the pharmaceutical value chain in three major ways:

1. As the growth of new biotechnology product launches continues to outpace that of traditional pharmaceutical products, research-based pharmaceutical companies have established significant access to these new technologies through collaboration and licensing agreements with biotechnology companies.
2. As the leading biotechnology companies have evolved through the development of key products, they have built up critical mass in the development and marketing functions to compete directly with integrated pharmaceutical companies across the value chain.
3. Biotechnology companies are key targets for M&A deals, as they provide growth potential by innovation.

Starting with the creation of Genentech in 1976 and the development of human recombinant insulin in 1978, biotechnology companies have played an increasingly important role in filling the R&D pipelines of pharmaceutical companies. According to Recombinant Capital (2005), more than 600 alliances are formed every year globally between pharmaceutical and biotechnology firms, with a deal values of over US$30 billion. The nature of these alliances varies: for instance, a biotechnology shop exchanges an exclusive license to market and sell a patented drug to a pharmaceutical company that is willing to pay some research costs upfront. Such agreements may also include limited use of the pharmaceutical company's manufacturing and distribution channels.

Table 3.1 Top best-selling products worldwide in 2014 (sales in US$ billion)

Rank	Product	Generic name	Sponsor	Sales	Technology
1	Humira	Adalimumab	AbbVie + Eisai	12.890	MAb
2	Solvadi	Sofosbuvir	Gilead Sciences	10.283	SMOL
3	Enbrel	Etanercept	Amgen + Pfizer + Takeda	8.915	Recombinant product
4	Remicade	Infliximab	J&J, Merck + Mitsubishi Tanabe	8.807	MAb
5	Lantus	Insulin glargine recombinant	Sanofi	8.428	Recombinant product
6	Rituxan	Rituximab	Roche	7.547	MAb
7	Seretide/ Advair	Fluticasone proprionat; Salmeterol xinafoate	GSK	7.058	SMOL
8	Avastin	Bevacizumab	Roche	7.018	MAb
9	Herceptin	Trastuzumab	Roche	6.863	MAb
10	Januvia/ Janumet	Sitagliptin phosphate	Merck	6.358	SMOL

SMOL small molecule, MAb monoclonal, antibody
Source: Evaluate (2015)

Table 3.2 Worldwide prescription drug sales from biotechnology products in 2020

#	Sponsor	2014 (US$ billion)	2020 (US$ billion)	CAGR (2014–2020) (%)	WW Market Share (2020) (%)
1	Roche	30.8	35.8	+3	12.9
2	Novo Nordisk	15.0	21.3	+6	7.7
3	Sanofi	16.0	20.8	+4	7.5
4	Amgen	17.6	20.2	+2	7.3
5	AbbVie	13.4	15.3	+2	5.5
6	Pfizer	10.8	14.0	+4	5.0
7	J&J	10.6	13.7	+4	4.9
8	BMS	3.7	12.8	+23	4.6
9	Merck & Co.	8.2	12.6	+7	4.6
10	Eli Lilly	6.0	12.0	+12	4.3

WW worldwide, CAGR compound annual growth rate, BMS Bristol Myers-Squibb
Source: Evaluate (2015)

Over the time, the negotiation power and balance in collaborations between pharmaceutical and biotechnology companies has shifted in favor of the biotechnology companies. Many young biotechnology ventures have become so-called 'born globals', small firms with global activities and thus many similar challenges like large firms (Gassmann and Keupp 2007). Some biotech companies have been negotiating successfully for co-promotion and manufacturing rights for current and future products with big pharmaceutical companies. For example, Exelixis (in collaboration with GlaxoSmithKline) retained North-American co-promotion rights for multiple compounds under mutual development.

Pharmaceutical companies may also make cash investments in exchange for a portion of future revenues and/or an equity stake in the biotechnology partner. Thus, it is not unusual for a large pharmaceutical company to have biotechnology holdings that give them a substantial piece of the action. For example, **Novartis** and **Morphosys**, a leading German biotechnology company in 2016, started their long-term partnership already in 2004 to discover and develop antibody-based biopharmaceuticals as therapeutic agents. In addition to usual service fees, Novartis made a 9 million euros investment in the start-up company by purchasing non-interest bearing convertible bonds of Morphosys. In 2007 the two companies signed a collaboration agreement to discover therapeutic antibodies, and in 2012 this strategic alliance was intensified by Morphosys' consent to provide new technology platforms to Novartis.

Research on alliances between biotechnology and pharmaceutical firms suggests that alliances are becoming more sophisticated and mature, that drug companies are central nodes in alliance networks, and that new biotechnology firms play a mediating role in transforming scientific knowledge into patented technologies (see Lin 2001).

The competition among biotechnology companies using the latest drug discovery techniques puts extra pressure on pharmaceutical companies to become technological leaders in their own respective areas. Traditional approaches and sequential experimentation in drug discovery have increasingly been complemented by automated, mass-production analysis of compound libraries and computer-based experimentation using several different new technologies. The complete mapping of the human genome is only further supplementing this paradigm shift in pharmaceutical R&D.

The application of new sciences and technologies resulted in the following essential changes in the drug discovery process (see also Nightingale 2000):

- The nature of scientific understanding of diseases becomes more fundamental. A more detailed understanding of drug target function in the context of a molecular mode of disease onset, progression and chronicity will increase the quality of applied targets;
- The scale of experimentation undergoes fundamental changes and shifts from an individual initiative to an automated mass-production process;
- The cycles of trial-and-error experimentation are complemented by computer simulations;
- Complementary screening is performed by computer simulations;
- Single compounds are replaced by compound libraries;
- Structural complexity and diversity of compound libraries increases.

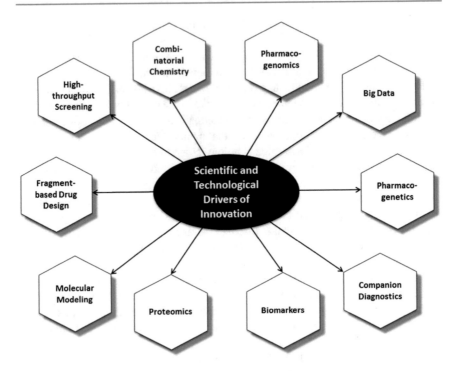

Fig. 3.3 New sciences and technologies as driver for innovation

This chapter focuses on new sciences and technologies as the underlying drivers for innovation in the pharmaceutical industry (Fig. 3.3). They enable the different parts of pharmaceutical R&D and lead to significant and revolutionizing results in pharmaceutical innovation.

3.2   The Challenge of Target Discovery

Target discovery is the first step in drug discovery and, thereby, targets need to be identified and validated. In an analysis of drugs approved by the FDA between 1980 and 2010, 435 effect-mediating drug targets were identified in the human genome, divided into the following target classes: receptors (193), enzymes (124), transporters (67) and others (51) (see Rask-Andersen et al. 2011).

From a process perspective, target discovery can be done following the molecular or the systems approach. While the molecular approach has been driven by technologies in molecular biology, such as genomics, the systems approach is related to the discovery of targets by studying diseases in whole organisms, such as animal models or clinical trial studies.

Tools for target identification are, e.g., reverse genetics screens, small-interfering RNAs (siRNAs) and gene expression profiling. All tools aim at correlating a change in a gene (genomics) or protein (proteomics) or genetic variation (genetics) with a human disease. For example, a genomics approach aims at identifying a new target by comparing normal and diseased tissues by gene expression profiling using gene arrays or proteomics to correlate the cellular function by measuring protein expression either with protein arrays or two-dimensional gel electrophoresis.

Target validation is the demonstration of the functional role of the target in a disease. In a first step, target validation is shown by the expression of the target in the diseased tissue, or if the modulation of the target expression has an impact on the disease phenotype in an animal model, such as in transgenic mice. Ultimately, target validation is given when the role of the target in a disease phenotype is proven in human studies.

For some infectious diseases, an open access drug target database, the TDR target database, has been set up. It comprises information from genome sequencing, functional genomics, protein structural data and drug targets of the disease-causing pathogens of malaria, tuberculosis, African trypanosomiasis, Chagas disease, leishmaniosis, or schistosomiasis, as well as data on the druggability of certain targets.

An alternative approach is pursued by the Accelerating Medicines Partnership (AMP), a collaboration of the FDA, the National Institute of Health (NIH), non-profit organizations and ten pharmaceutical companies that provided US$230 million for the search of new targets to treat Alzheimer's disease, type 2 diabetes, rheumatoid arthritis and lupus. This public-private partnership aims at transforming the current R&D model for developing new diagnostics and drugs by jointly identifying and validating promising biological targets to reduce the time and cost of doing R&D. Not all major research-based pharmaceutical companies have joined AMP: For instance, Amgen, AstraZeneca, Bayer Healthcare, Novartis and Roche did not see the potential of precompetitive joint research activities. AstraZeneca prefers focused research collaborations in target discovery, and Amgen acquired Islandic Decode Genetics in 2012 for US$415 million to increase Amgen's capabilities to develop better targeted drugs.

Functional genomics as a method in drug discovery denotes when diverse technology platforms including molecular biology, cell biology, proteomics, bioinformatics and model organism are used to generate multi-dimensional profiles of genes to identify promising new drug targets (Kramer and Coher 2004). This method also includes the analysis and interpretation of large sets of data, and the translation of the resulting information into understanding the role of genes in biological pathways and related diseases.

Natural occurring mutations in humans can also be used to evaluate efficacy and safety aspects related with drug targets, especially if preclinical animal models have limited informative value. The thinking behind this approach, also named "experiment of nature", is that data derived from patients can be used to correlate drug targets with diseases. The results of the experiment of nature can be used to mimic the effect of drug target modulation and to provide pharmacodynamic and pharmacokinetic information before the start of clinical trials. The effectiveness of several of these preclinical models for target validation is outlined in Table 3.3.

Table 3.3 Human relevance of preclinical models for target validation

Model	Human relevance	Causality in humans
Cellular models	Ineffective	Ineffective
Animal models	Effective, with some limitations	Ineffective
Human epidemiology	Highly effective	Ineffective
In vivo expression studies	Highly effective	Ineffective
Natural conditions	Highly effective	Highly effective
Human genetics	Highly effective	Effective, with some limitations

Source: Plenge et al. (2013)

Finally, pharmacophylogenomics is used in drug discovery to improve target selection and validation. It is "the application to genomics of prin-iples and techniques from evolutionary biology, to achieve a better understanding of gene function" (Searls 2003).

Today, large sets of data from genomics, proteomics, transcriptomics, or metabolomics have resulted in a paradigm shift towards complex gene networks, a network-based view of diseases and ever-more complex drug discovery strategies.

3.3   High-Throughput Screening: Fail Earlier, Succeed Sooner

As molecular biology has revolutionized drug research in the past 30 years, drug discovery shifted from animal studies-focused to target-oriented research. This new target identification and validation paved the way for high-throughput screening (HTS) technologies. The entry into drug discovery is usually the identification of low-molecular-weight modulators with a specific activity in a biochemical target assay. Such hits can be identified via various strategies such as knowledge-based approaches, combinatorial chemistry and HTS (Bleicher et al. 2003). HTS is the most used technology in discovery research to provide hits for lead finding.

HTS is the (biochemical) screening of a large number of chemical compounds to identify hits in an in vitro assay, usually performed robotically in 1536 or 3456-well microtiter plates. It comprises of a system for data handling, an array of compounds to be tested, a robot to perform the testing and a biological test configured for automation. Pharmaceutical companies have invested significantly in the HTS technology and their screening collections to build a screening platform for diverse drug targets. The size of corporate screening collections has been increased significantly from 200,000–300,000 in 2000 to around 2,000,000 today. As corporate screening collections have few overlaps in terms of compound identity and similarity, Kogej et al. (2013) proposed to increase the value of screening collections by exchanging corporate collections of pharmaceutical companies to expand the chemical space that can be accessed by HTS. Importantly, typical HTS runs comprise 10^5–10^7 compounds. If contrasted to the theoretical possible number of compounds in the chemical space of 10^{40}–10^{100}, the results of HTS runs are in favor of this technology (Macarron 2006).

Along with the sequencing of the human genome, bioinformatics, and combinatorial chemistry, HTS was anticipated to deliver a wave of new drug targets and therapies with lower costs and reduced timelines. However, the very high expectations have only been fulfilled in part. HTS has been criticized as a technology that provides data of poor quality, is expensive and time-consuming and it fails to provide lead structures (Macarron et al. 2011). It has also been blamed to contribute to the high attrition rates and low productivity of pharmaceutical R&D, as the phenotypic screening approach resulted in the development of more first-in-class drugs (28) approved by the FDA between 1999 and 2008 than the target-based approach (17) (Swinney and Anthony 2011). Several factors contributed to the lower success rate of the target-based approach, such as poorly validated targets, early non-drug-like compound libraries, small screening libraries, artificial assay systems, limited bioinformatics capacities, and a lack of appropriate animal models. The drawback of phenotypic screening is the time-consuming identification of the drug target after compounds have been identified to be active in a phenotypic screen.

Hence, the target-based approach to HTS is a useful strategy in some cases but certainly not applicable for all disease mechanisms. It is more expensive than alternative approaches, but considering the high number of compounds screened and the data provided, it is a comparably cheap technology. Although the upfront investments for HTS units are significant, the costs for a single HTS are around US$75,000 and thus only 10–20% higher than for other methods (Macarron et al. 2011). The success rates of alternative methods, such as fragment-based screening, structure-based design or virtual screening, are not higher. HTS, however, is applicable to more targets, and less initial knowledge about the target and ligand is required.

The success rates of HTS in identifying hits have been analyzed for **GlaxoSmithKline** (GSK). In Macarron's (2006) study of 42 HTS-laboratories, the average success rate was 56% ranging from 8 to 100% (depending on the target class). In later work, including analysis of Pfizer, GSK, Novartis and Sanofi, Macarron et al. (2011) showed that the proportion of leads derived from HTS was in the range of 48–84%.

Even when hits are identified as a result of HTS screening, they are not always useful for further use in medicinal chemistry. Consequently, HTS is often used in parallel with other approaches. Hit identification and lead optimization are more and more done on basis of cellular assays rather than biochemical assays, as cellular assays are more physiologically relevant.

In addition to lead finding, the major impact of HTS for drug discovery was the technology push related with HTS. Research-based pharmaceutical companies invested significantly in the underlying technologies ranging from compound collections, to compound management, assays technologies, bioinformatics to funding of technology-providing biotechnology companies. Technologies, that were originally developed for HTS and hit identification, are now also used in other drug research areas, for example in high-throughput pharmacokinetics.

Combinatorial Chemistry: Cut Experimental Cycle Times

Along with HTS, combinatorial chemistry is considered an essential tool for drug discovery. During its industry-wide implementation, both technology platforms accounted for more than half of all spending on new discovery technologies in the pharmaceutical industry (Reuters 2002).

The emergence of combinatorial chemistry was triggered by the increasing application of HTS technologies. With the surge of HTS technologies, a bottleneck in the discovery process occurred as the production of compounds did not expand at the same rate. As rapid progress in HTS allowed the screening of tens-of-thousands of compounds rather than just a few hundreds, it became obvious to pharmaceutical companies that they could now test all their compounds very quickly. Thus, improvements in screening technologies increased demand for compounds and created a 'reverse salient' in synthetic chemistry (Nightingale 2000).

Combinatorial chemistry is a synthetic technology to generate compound libraries rather than single compounds by synthesizing all possible combinations of a set of small chemical structures or building blocks. Hence, combinatorial chemistry is a mass-production technology that synthesizes large numbers of compounds in parallel for the use in HTS. When introduced in the 1990s, combinatorial chemistry permitted experimental cycle time reductions by a factor of 800, and lowered costs and risks by a factor of 600, compared to traditional methods (Booz Allen and Hamilton 1997).

The compound libraries of pharmaceutical companies typically include historical compound collections, natural products and combinatorial compound libraries. While at the beginning of HTS and combinatorial chemistry compounds were generated just to increase the size of the libraries without account for target-related design or filtering criteria, the compound libraries are eliminated from these standard screening sets and include now smaller and focused drug-like subsets to increase the success rates of HTS.

Hit and Lead Generation Technologies Beyond HTS

Target-based screening was initially implemented to improve the drug-like properties and selectivity of new drugs. Swinney and Anthony (2011) revealed a negative impact of HTS and the related target-based approach to the drug discovery success rates and R&D productivity. This is in line with previous reports that some of the most specific drugs are directed towards a drug target that does not play a central role in disease pathology, and that many drugs that were anticipated to interact with one specific target have unwanted off-target side-effects (Greef and McBurney 2005). This has led to a discussion about how drug discovery should be performed, target-specific or systemic:

- The system-based approach is a hypothesis-agnostic assay, an approach that is based on phenotypic changes seen in vitro or in vivo. Phenotypic screening is the testing of a large number of compounds in a systems-based approach.

- The chemocentric approach is a drug discovery approach that is based around a specific compound or a compound class.
- The target-based approach is a hypothesis-driven method that aims to manipulate a biological system by pharmacologically modulating a specific target.

A recent analysis by Eder et al. (2014) of 113 first-in-class drugs approved by the FDA between 1999 and 2013 showed that most of these drugs (78) were developed by target-based screening. 33 drugs were developed without knowing the drug target, of which 25 by chemocentric approaches and 8 by phenotypic screening. The target-based drugs could be grouped into small molecule drugs (45) and biologics (33). The biologics were typically identified by screening (antibodies) or rational design. The 45 small molecule drugs resulted from HTS (18), chemocentric approaches (18), rational design (6), fragment-based screening (1), *in-silico*-screening (1), and low-throughput-screening (1). HTS has thus played a prominent role in drug discovery, but other technologies are also important in the provision of hit and lead structures.

Fragment-based drug design is one of these alternatives for drug discovery. In contrast to the conventional screening methods that aim at analyzing as many potent hit and lead compounds as possible, fragment based-drug design is based on screening of smaller numbers of compounds in the hope of finding low-affinity fragments. It utilizes the principle of the fragmentation of a lead compound into smaller parts (usually with diameters of less than 250 Da), the optimization of single functional groups by computer algorithms, and NMR spectroscopy or X-ray crystallography. The optimization of each interaction of the lead in the binding site and subsequent incorporation of all interaction sites into one drug molecule provides compounds with better drug-like properties than the hits/leads provided by HTS. Screening of a fragment library with 10,000 compounds comprises more chemical diversity than a typical HTS screen. An analysis of 45 screens performed by Abbott, Hajduk and Greer (2007) showed that the success rate of the fragment-based approach was 76%, which led to the conclusion that the fragment-based approach can result in a higher hit rate than HTS. In practice, both approaches are used complementary. While HTS is providing fast result and initial leads, fragment-based screening takes more time but provides higher quality lead structures.

While in the target-based approach hits and leads are identified and optimized based on their activity towards a specific target, Butcher (2005) outlines a cell systems biology approach, placing biology at the beginning of the process. Primary human cell systems are used to model aspects of the disease biology and to identify and optimize leads. The drug candidate itself, not the drug target, is used for validation. In the same vein, systems pathology and systems pharmacology have been proposed as alternative paths in drug discovery (Greef and McBurney 2005).

Another trend is the use of induced pluripotent stem cells (iPSCs) for drug screening. iPSCs can be used to produce patient-specific cell lines as models of specific disorders. The iPSC-derived cells help to understand the disease pathogenesis, support target identification and validation as well as they are useful for HTS screens. Takeda Pharmaceuticals is one of the protagonists in this field. Collaborating with Kyoto University's Center for iPS Cell Research Application

(CiRA), the Japanese pharma giant developed clinical applications of iPSCs (for details see Chap. 6). Numerous other pharmaceutical companies have also opened their discovery research organizations for new collaborative models to increase the success rates of their lead finding units. Examples are the open innovation models of Eli Lilly (PD^2), Bayer Healthcare (G4T) or the Innovative Medicines Initiative European Lead Factory. Likewise, industry-academia or industry-industry collaborations are used to complement know-how and capabilities in drug discovery, e.g. MedChemica or the Bayer Healthcare and AstraZeneca collaboration for the exchange of screening capabilities.

3.4   The Bioinformatics and eHealthcare Revolution

Advances in molecular biology, genomics, proteomics and other new technologies made it necessary for biotechnology and pharmaceutical companies to deal with huge amounts of biological and chemical data. Information from the human genome, gene expression profiles, protein-protein interactions and metabolic pathways are used to understand the mechanisms of diseases and to discover new drugs.

Bioinformatics, eBiology, dry biology, *in silico* biology, computational biology, conceptual biology or knowledge discovery are all used as synonymous terms. Bioinformatics integrates and applies data from diverse sources of drug discovery and clinical development in a systemic way for the identification, ranking and progression of drug discovery programs. The role of bioinformatics in drug discovery can only become more important, as the worldwide amount of knowledge doubles every seven years (Davis and Botkin 1994).

Pharmaceutical research is among the leaders in knowledge production. Drivers for the increasing knowledge accumulation include the escalating usage of novel drug discovery technologies as well as external knowledge acquisition. Genomics data, such as data from sequence analyses, genetic variations, target similarity data, data from target functions, expression profiles, knockouts, screening data, clinical data and data from pharmacogenomics are all integrated in the analyses to address specific questions in addition to the "wet-lab" experiments. New technologies in drug discovery produce huge amounts of data and information. Given that a single pharmaceutical lab can generate hundreds of gigabytes of data per day, sophisticated information technologies and algorithms are needed to scan and interpret them.

Computational systems biology aims to integrate experimental biology research and computational research to better model and understand the complexity of biology (Kitano 2002). Computational biology has two distinct roles: (1) knowledge discovery or data mining and (2) simulation-based analysis. While the first part aims at extracting pattern from a huge set of data, the latter tests scientific hypotheses in *in silico* experiments. Combinatorial informatics or chemogenomics is another approach in drug discovery that combines *in silico* experiments with wet-lab data (Agrafiotis et al. 2002).

In addition to corporate information, biotechnology and pharmaceutical companies also access data from public databases, such as the Kyoto Encyclopedia of Genes and Genome, the Alliance for Cellular Signaling, PubChem, DrugBank, Chemical Entities of Biological Interest or ChemBank. These openly accessible databases are the foundation of pre-competitive research and the basis for public-private partnerships.

Another initiative is the integration of structure-activity relationship (SAR) data from various sources into one single system, including around 4.8 million compounds and approximately 600,000 SARs for 3000 protein targets from published and private sources, to get a better understanding of the pharmacological space (Paolini et al. 2006). The Similarity Ensemble Approach can be used to search compound databases and to build cross-target similarity maps. The tool groups receptors by chemical similarity of their ligands to help identify unknown relations between receptors and ligands.

Bioinformatics tackles the challenge of multiple and inconsistent sources of information (inside and outside the companies), the fragmentation of data, and the integration of information into one coherent entity. As data sources expand, robust data analysis and knowledge discovery becomes more important. Even small start-up companies operate relatively globally in R&D: They acquire their knowledge through cooperation with other complementary companies as well as through public databases (e.g., Internet).

For instance, researches at **Johnson & Johnson** (J&J) developed an integrated bioinformatics platform named ABCD that provides a framework for organizing chemical and biological data, one interface to access data derived from multiple sources and a framework to support the project team-based structure of J&J's R&D organization. The benefit is to have all relevant data available from one single system (Agrafiotis et al. 2007).

Information technologies in pharmaceutical research must deal with both the management of data and knowledge within the corporate boundaries and the linkages to the outside research community. Baumann (2003) summarizes the major tasks of information technologies in pharmaceutical R&D as follows:

- To provide and manage databases for the tremendous amount of information;
- To allow the generation of compound profiles for improved target identification and screening;
- To manage genome and protein sequences;
- To visualize 3D data;
- To collect data on model organisms;
- To manage the huge amount of data from the clinical tests and provide feedback to the early phases of drug discovery;
- To enable accessibility and sharing of knowledge within the corporation as well as to outside collaborators.

One of the foremost tasks of information technologies in pharmaceutical research is to handle the large amounts of complex data generated throughout all phases of the R&D process. For example, the U.S. Trials Register has become a global source for industry-sponsored phase II–IV clinical trials and an option to access big quantities of data from ongoing clinical trials.

Furthermore, electronic health databases have become an important source for pharmacoepidemiological and translational research. By using these databases, researchers obtain knowledge on the short- and long-term effects of drugs, identify responders and non-responders as well as patients at risk for adverse effects or access information on the use of a drug across different geographies.

Already a key communications and information tool, the internet will provide pharmaceutical companies with the ability to more effectively interact with partners, regulators and consumers. Internet-based technologies hold the potential to impact every stage of the pharmaceutical value chain. The ability to access and share data, and to interact and communicate within and across organizations, will determine how successfully new technologies are integrated. Investment in internet technologies should realize significant productivity gains across the research and the development functions.

Thorough and pervasive harnessing information, for example in respect to a specific therapeutical area or a disease, can be a competitive edge for a pharmaceutical company. Large amounts of data are available for neurogenerative diseases, such as Alzheimer's disease. The Alzheimer's Disease Neuroimaging Initiative, Brain Research through Advancing Innovative Neurotechnologies, the Human Brain Initiative, the Allen Brain Initiative or the Blue Brain Project have begun to use imaging, genomics and computer technologies to decode the functional, structural and biochemical networks of the brain. Consumer genomics and electronic medical reports systems are another source of information and basis for the correlation of systems-biology derived information with adverse events, co-morbities and drug responses that are stored in electronic medical reports or patients' questionnaires. The discovery of disease mechanisms, the identification of predictive mechanistic models and the provision of novel drug targets based on this knowledge are potential starting points for next generation of therapeutic advances.

However, there is still skepticism on the return-on-investment (ROI) of big data, and critical issues such as an efficient process for data sharing or common policies for data exchange, have not been solved so far (see e.g., Manji et al. 2014). Barriers to the adoption of new internet technologies also include concerns over the security of company proprietary information, regulatory concerns about patient confidentiality.

"Big Data" and Its Relevance for Pharmaceutical R&D[1]

The successes of companies such as Facebook, Amazon and eBay have raised the expectation that pharmaceutical companies can also generate value out of big data by somehow adopting new business models. While Facebook and its peers have easy

---

[1]With thanks to Dr. Lars Greifenberg, Director R&D IT, AbbVie Deutschland.

access to high quality consumer data, as they compile and own the data with their general terms and conditions, pharmaceutical companies struggle to access relevant data for their core business, i.e., developing drugs. The main source for reliable patient data is the proprietary information from their own clinical trials. Data generated by health insurances, social media or public sources often lack granularity and quality.

Missing data standards and the absence of controlled vocabularies prevent the compilation of large and reliable data volumes—the prerequisite for detailed data analysis. However, the introduction of electronic health records (EHR) and the trend to capture more and more molecular data in standard healthcare (such as genetics, proteomics, metabolomics) are an opportunity to access more relevant patient data in the future. Data security and privacy concerns as well as missing legal regulations for data abuse still prevent pharmaceutical companies from accessing new data sources in many countries. It is common sense that patient data need to be protected and anonymized before being shared and used by a second party. Unfortunately, genetic information represents a personal fingerprint and cannot be anonymized. The creation of a safe environment in which the individual sees clear personal and collective benefits if data is shared seems to be the only way to realize this vision.

If EHRs are made available for pharmaceutical companies, this would allow scientists to better understand the underlying mechanism of diseases, to better design clinical trials, to increase the probability that patients respond to new treatments positively, and to reduce adverse events. Thus, big data not only has the potential to increase R&D efficiency, if it included also data from healthy humans (pre-disease), it would also allow to change the industry's business model from curing to preventing diseases.

3.5 Proteomics and the '-omics' World

The term proteome refers to all the proteins expressed by a genome, and proteomics is concerned with proteins produced by cells and organisms. The approximately 30,000 genes defined by the Human Genome Project translate into several hundred thousand proteins when alternate splicing and post-translational modifications are considered. While a genome largely remains unchanged, the proteins in a cell change dramatically as genes are turned on and off in response to the environment.

Most drugs work on proteins or protein receptors. Hence, a key challenge of proteomics is to identify differences between the pattern of a healthy and a sick person, analyze them, and then identify and isolate target proteins. Consequently, proteomics covers the effort to obtain complete descriptions of the gene products in a cell or organism. The goal of proteomics is to analyze the information flow within a cell and the organism through protein pathways and networks. Thus, proteomics includes not only the identification and quantification of proteins, but also the determination of their localization, modifications, interactions, activities, and, ultimately, their function.

It is generally believed that proteomics will eventually help us understand diseases and their underlying mechanisms, and further facilitate the target-based approach in drug discovery. In addition, proteomics technologies are applied in clinical development, as for example in cancer, to detect cancer earlier or to identify new drug targets and biomarkers.

Other 'omics' technologies emerged in the past years to enable modern drug discovery. For example, metabolomics is the study of in vivo metabolic profiles to provide information on drug toxicity, diseases mechanisms or gene functions (Nicholson et al. 2002). Here, lipidomics is the systems-level analysis and characterization of lipids and their interacting moieties (Wenk 2005). Toxicogenomics is the application of gene expression technologies to toxicology to early identify hazards and to reduce risks associated with drug candidates (Ulrich and Friend 2002).

3.6   Pharmacogenetics and Pharmacogenomics

The response of a patient to a drug depends on intrinsic factors, such as health status, gender or genes, and extrinsic factors, such as diet or the use of concomitant medication, both influencing pharmacokinetics and pharmacodynamics of a drug. The impact of genes on drug response started to be analyzed with the sequencing of the human genome. Today, we know that the individual response to a drug depends on the following factors:

- Genes relevant for the absorption, distribution, metabolism and excretion of the drug;
- Genes encoding drug targets; and
- Genes affecting disease progression.

All factors together influence the pharmacokinetics and pharmacodynamics of a drug, and thus the diverse outcome of clinical trials in different subpopulations of patients.

Pharmacogenetics is the study of inherited genetic differences in drug metabolic pathways which can affect individual responses to drugs, both in terms of therapeutic effect as well as adverse effects. Pharmacogenetics is based on clinical observations of the efficacy, safety and tolerability of drugs. The inter-personal differences of drug response can be associated with specific biological markers that in turn allow to forecast individual drug responses. By analyzing different drug responses of different individuals with the same medication, it is possible to fine-tune medication for individual patients. Patients are divided into two sub-populations, the responders and the non-responders, based on whether they respond to a drug better than the average patient. Better drug efficacy and better patient tolerability is the outcome. Pharmacogenetics thus helps decrease attrition rates in R&D and reduces adverse events after the new drug is commercialized.

While genomic technologies permit a better understanding of drug target function in genomic population subsets and even individuals, they raise great commercial and financial concerns. Much time and resources must be spent to develop genetic profiles, while market sizes for tailored drugs are much smaller, requiring very different portfolio management strategies. Exploiting this opportunity will require that companies leverage genotype-based diagnostics into personalized medicine, completely shifting the end-game equation from high-volume/high-value (i.e., blockbuster) drugs to small-volume/higher-value (i.e., individualized) drugs. While some believe that personalized medicine will be of limited importance, others are convinced that it will have a broad impact in the industry, and that genotype-based elimination/exclusion of major side effects will eventually create even larger blockbuster products than is presently possible.

If pharmacogenetics manages to perfect the stratification of patients into responder and non-responder sub-segments, it has the potential to change pharmaceutical R&D and its output significantly:

- Understand why some drugs lack efficacy in certain patient segments;
- Justify the incidence of adverse drug reactions;
- Explain the variability of drug response during clinical trials; and
- Help to design better clinical trials.

One key technology in pharmacogenetics is DNA sequencing. Technological advances have brought down the cost for DNA sequencing substantially. Today, is possible to sequence a human genome (3.27 billion base pairs) for around US$10,000. Sequencing the bacterial genome of *Chlamydia pneumonia* (a mere 1.2 million base pairs) cost still more than US$1 million in 2000. In consequence, genome sequencing has developed into a cost-efficient method that enables pharmacogenetics and thus personalized medicine.

Biotechnology companies such as Illumina or Life Technologies are specialized in providing new sequencing technologies for the use in pharmacogenetics or pharmacogenomics. Both companies recently invested in third generation sequencing technologies (3GS) to further increase sequencing throughput at reduced costs. For example, Illumina acquired GeneLogics LifeScience Software in 2015 and Conexio Genomics in 2016. DNA sequencing expertise will become even more important: In 2017, the global sequencing market is forecast to reach US$4.27 billion, 39% of which for healthcare applications (Mohamed and Syed 2013).

The European Medicines Evaluation Agency (EMEA), the U.S. Food and Drug Administration (FDA) and the Japanese Pharmaceuticals and Medical Devices Agency (PMDA) have already promulgated guidelines on the role of pharmacogenetics in pharmaceutical R&D, underscoring the importance pharmacogenetics already plays today. However, many challenges remain, such as linking genetics variations to a specific disease and a relevant clinical outcome, or privacy issues of patients involved in genomic studies. All this removes the vision of truly personalized medicine further out into the future.

As it was the case for pharmacogenetics, new genomic sequencing technologies were also the start of pharmacogenomics. Pharmacogenomics is the genome-wide systemic assessment of effects of drugs on gene expression. It does not analyze inter-individual differences with respect to the effects of a drug, but analyzes the differences of various drugs on the expression profile of certain tissues. Today, pharmacogenomics is used in lead finding to identify the best drug candidates. It is used in in vitro pharmacology profiling, i.e., the screening of compounds against a broad range of drug targets (off-targets) that are distinct from the primary therapeutic target(s), which helps identify specific interactions that may be the cause for adverse drug reactions. Off-target effects are usually harmful, but in some cases these properties have induced important positive effects resulting in serendipity findings for the use of a drug in new therapeutic indications. Thus, sharing of side-effect profiles can be used for target discovery. It can also help avoid dose-related side-effects to certain drugs with known genomic biomarkers.

Biomarkers are naturally occurring genes, molecules, or other characteristics that can be used as indicators of normal biological processes, pathogenic processes, or pharmacological responses to a therapeutic intervention. For instance, biomarkers help measure the delivery of drugs to their primary targets, understand the predicted pathophysiology and how it is altered, and monitor variables of clinical relevance. Biomarkers are useful in saving R&D resources by enabling early proof-of-concept studies.

Overall, the number of biomarkers used in drug discovery increased significantly over the past years, although the number of biomarkers accepted as surrogate end points in pivotal clinical trials has remained low (Hurko and Jones 2011). A surrogate endpoint is defined as a biomarker intended to substitute a clinical end-point. For example, lowered blood pressure is an accepted surrogate endpoint for stroke or myocardial infarction.

Predictive biomarkers utilize a baseline characteristic of a patient to predict whether the patient will benefit from a therapy at all (Beckman et al. 2011). For example, the increased understanding of cancer biology has resulted in the knowledge that cancer is a heterogeneous disease and needs to be treated by specific cancer drugs. Personalized drugs have been developed specifically against a defect in cellular signaling that occurs only in particular patient sub-populations. The defect and thus the sub-population are characterized by the incidence of a discrete biomarker. Prominent examples of personalized drugs are Herceptin (Roche) for the treatment of breast cancer characterized by HER2 overexpression or Gleevec (Novartis) for the treatment of chronic myeloid leukemia in patients typified by the BCR-ABL fusion gene. The identification of patients that benefit from such treatment is usually conducted with companion diagnostics, which are in vitro diagnostics assays that provide the information that is essential for the safe and effective use of the therapeutic intervention.

Today, the growing importance of biomarkers in drug R&D has been recognized by relevant regulatory authorities (see HHS and FDA 2005, or FDA 2014). In addition, the Biomarker Consortium has been founded in 2006 by the FDA in collaboration with the National Institute of Health (NIH) and the Pharmaceutical Research and Manufacturers of America (PhRMA) to facilitating the development of personalized medicine by supporting research on biomarker-based technologies and medicines, and developing therapies in the fields of cancer, neurodegenerative diseases, metabolic disorders and inflammatory diseases. In 2010, the Biomarker Consortium launched a pioneering adaptive clinical trial with numerous drug candidates to treat breast cancer. The goal is to identify which biomarker subtype benefits most from which therapeutic intervention. The results are expected to make subsequent phase III trials more successful.

In additional to the technical and regulatory advantages, the provision of companion diagnostics can be an important decision criterion for pricing and reimbursement of a new drug. For example, the prices of novel cancer drugs could reach US$100,000 per year and patient. Companion diagnostics help justify the price and assure that only responders are treated with the drug. Companion diagnostics are regarded as a value-creating and science-driven approach to drug development, as they lower the risk of clinical trials, increase the speed to market and reduce the overall development costs.

Pharmacogenetics and pharmacogenomics technologies are cornerstones for personalized healthcare (PHC). Its major benefits are:

- More powerful medicines;
- Better, safer drugs the first time;
- More accurate methods of determining appropriate drug dosages;
- Advanced screening for disease;
- Improvements in the drug discovery and approval process;
- Better negotiating position for pricing and reimbursement.

Finally, both technologies are expected to lead to an overall reduction in the costs of healthcare due to decreases in:

- The number of adverse drug reactions;
- The number of failed drug trials;
- The time it takes to get a drug approved;
- The length of time that patients are on medication;
- The number of medications patients take to find an effective therapy;
- The effects of a disease on the body (through early detection).

3.7 Computer-Based Drug Discovery

Methods of computer-based drug design can be applied at different phases of drug discovery, such as virtual screening for hit identification, fragment-based drug design in lead finding or molecular modeling in lead optimization.

Virtual screening has become the most important alternative to HTS by screening large compound databases *in silico*, selecting a limited number of drug hits for biological testing and identifying novel chemical entities that have the desired properties and are most likely to bind the drug target.

X-ray crystallography and nuclear magnetic resonance (NMR) are already used for protein structure determination at the atomic level. Knowledge of the three-dimensional structure of a target protein is the starting point for structure-based approaches to drug design by defining the features of the complementary structures of ligand and target. The respective information is used to optimize drug molecules by building more effective interactions with the target protein, resulting in potentially improved potency and higher selectivity. The three-dimensional structures of proteins are determined by high-throughput technologies, thus enabling e.g. fragment-based drug design.

While the novel screening methods are helping to find the relevant substances by eliminating the irrelevant substances, molecular modeling derives the design of the target molecules analytically. This approach seems to be far more effective and efficient than screening methods because it is based on an analytical process rather than serendipity.

Most drugs work through an interaction with the target molecule or protein, which causes the respective disease. The drug molecule inserts itself into a functionally important crevice of the target protein, like a key in a lock. The drug molecule is then connected to the target and either induces or, more commonly, inhibits the protein's normal function.

Consequently, a better and more direct understanding of the drug-target interaction would make screening of hundreds of thousands of substances obsolete. If it were possible to identify the appropriate target for a given therapeutic need in advance (including the structure of the target protein), the structure of an ideal drug molecule could easily be designed to interact with the respective target.

Hence, molecular modeling is involved in exploiting three-dimensional structures of molecular targets as well as the respective drug molecules. After determining the three-dimensional atomic architecture of the target protein and its functionally critical regions, a variety of specialized programs on interactive graphics workstations come into play. A design team develops and evaluates ideas for structures of drug molecules that complement the unique structure and electronic environment of the target protein. The medicinal chemists then chemically synthesize the most promising candidate structures. As in conventional drug discovery strategies, biochemists measure the ability of these newly synthesized drug candidates to produce the intended effect upon the target protein. Crystallographers then re-determine the structure of the protein target, but now in combination with the candidate drug molecules. They see the detailed structural interactions achieved by the candidate drugs with their target. The scientists relate the performance of such compounds measured by familiar biochemical techniques to its structural interactions with the target as revealed by crystallography. The design team then incorporates the results of this analysis into the next generation of compounds.

This drug design methodology consists of iterative cycles of design, simulation, synthesis, structural assessment, and redesign. The pharmaceutical industry tries to adapt design rules known from the machinery industry to the far more complex world of molecules.

3.8   Conclusions

Since its establishment, the pharmaceutical industry has always been one of the most research-intensive and innovative sectors of manufacturing. The scientific and technological revolution in the pharmaceutical industry has been driven by biotechnology and strongly influenced drug discovery in the past two decades. Today, complex sciences and technologies such as HTS, combinatorial chemistry, genomics-based technologies, proteomics or rational drug design, are used to identify hits and to discover lead substances. The novelty, complexity and strategic impact of these technologies as well as the sequencing of the human genome have led to the general opinion that a new wave of breakthrough innovation is to come.

Besides biotechnology, the pharmaceutical industry is also affected by the rise of big data. This is nothing new to the pharmaceutical industry since it was always dealing with big data. But the combination of personal data with social data, and the use of data generated by smart wearables might have an even higher impact on life sciences soon. The challenge for pharmaceutical companies is new: many of its big data competitors are not part of the regulated world of the FDA and other regulatory institutions. They come from the consumer electronic and software world: Apple, Google and many e-health platforms are providing platforms for end consumers generating huge amounts of data. Digital therapies for obese children will use their smart phones, and as social media technologies are much more pervasively accepted, they might have a bigger impact than traditional pharmaceutical drugs. The challenge for traditional pharmaceutical companies is bigger than the integration of new technologies caused by the emergence of the biotechnological thinking. The consumer/end user will decide on the way of value creation. It is all but certain that software-driven platform providers and consumer electronics companies will capture large parts of the created value. New business models and new competencies are required to manage this challenge. Pharmaceutical companies will face an innovation race with fast acting and mostly unregulated competitors. Instead of traditional regulation of drug discovery, new rules on data protection and privacy apply. Not a single pharmaceutical company of today has the competence and experience to compete successfully in the new globalized healthcare business.

The even bigger challenge is to manage and influence the ecosystem of the new healthcare system and participate in the value creation and value capturing process with the right business models and collaboration partners. Considering the time which has been needed for including and accepting biotechnology into the pharmaceutical innovation pipeline, this challenge is immense: Nothing less than the entire world of software, social ecosystems, platforms and new business models must be integrated into the pharmaceutical innovation pipeline.

The Pipeline Challenge: How to Organize Innovation

4

"*Our industry is poised to translate our most promising scientific breakthroughs into meaningful treatments capable of tackling the most urgent and vexing medical challenges of our times.*"

Kenneth C. Frazier,
Chairman & CEO, Merck & Co.

4.1 The Relevance of Pipeline Management

As in many other industries, innovation is the key driver of sustainable growth, and thus comes the challenge to organize innovation more effectively and efficiently than competitors. Somewhat specific to the pharmaceutical industry, however, is the lasting presence of risk throughout all stages of R&D and even beyond: These risks can be split into technological, market and commercial risks.

Technological risks consist of the scientific and technical uncertainties of discovering and developing a new drug, and showing that this drug is efficacious and safe in vitro and in vivo in animals and humans. Market risks are related to strategic uncertainties, such as sufficient patient numbers to be treated at the time point of market launch and thereafter, or how many competitors are likely to provide similar products. Commercial risks are execution uncertainties related to pricing and reimbursement.

The fundamental uncertainties of scientific and technological risks in pharmaceutical R&D are enormous and result in an overall likelihood of technical and regulatory success (PTRS) for drug R&D of only 4%. Most of the R&D projects fail and influence the total costs of R&D negatively. Worse even, pharmaceutical managers also need to factor in the exceptionally long project durations of 14 years on average per R&D project. The related capitalized costs can easily accumulate to more than US$2 billion.

If pharmaceutical companies want to survive, they must develop and feed new products to target markets at a predictable and steady rate. In consequence, the drug

© Springer International Publishing AG, part of Springer Nature 2018
O. Gassmann et al., *Leading Pharmaceutical Innovation*,
https://doi.org/10.1007/978-3-319-66833-8_4

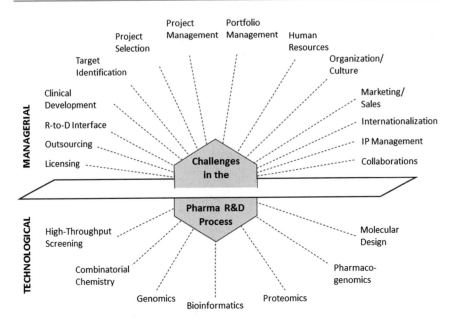

Fig. 4.1 Challenging aspects in the pharmaceutical R&D process

discovery and innovation process must somehow manage to overcome the managerial and technical uncertainties successfully—a highly complex undertaking given the long time horizons involved (Fig. 4.1).

To cope with the technical aspects, pharmaceutical companies have relied on increasing investment in R&D over the past years and decades (see Fig. 4.2). In turn, this commitment further increased the pressure on R&D managers to make well-considered investment decisions in R&D.

Complexity and Phases of the R&D Process

Blockbusters seem to get all the attention in the pharmaceutical industry, but one must consider that there is one of the most complex R&D systems across all industries working behind the scenes. Although pharmaceutical companies are among the top investors in R&D worldwide, they are, on average, only the sixth most effective industry in generating innovation, behind aerospace and defense, automotive, electrical/electronics, chemicals and IT hardware (Reuters 2002). This discrepancy requires an explanation: Why do high R&D investment with stringent R&D processes not correlate with a high R&D output?

Compared to other industries, the innovation process in the pharmaceutical industry has some unique characteristics, most notably the very stringent regulatory environment, which has a direct impact on the phase model of R&D, its timing, the PTRS and related costs. In general, only 1 out of 5000–10,000 compounds analyzed

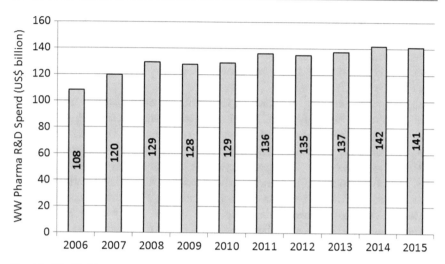

Fig. 4.2 Worldwide total R&D expenditure, by year (2006–2015). Source: Evaluate (2015)

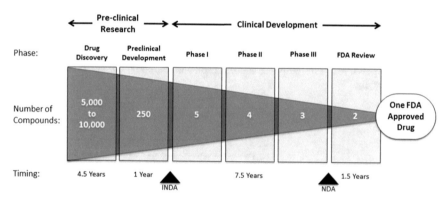

Fig. 4.3 The pharmaceutical R&D process. *INDA* investigational new drug application, *NDA* new drug application, *FDA* Food and Drug Administration

in drug discovery make it through the R&D process, get approved and are launched to the market (see Fig. 4.3): a very small output, in particular in view of the high-throughput technologies that are applied in drug discovery. Perhaps even more notable, the average new drug does not generate enough total sales to recoup its costs of R&D.

Other specificities are: While in most industries the decision to terminate a project is made based on economic considerations, the typical reasons in the pharmaceutical industry, at least in drug discovery, are primarily scientific or technological.

Additionally, the long timelines of more than ten years from initial idea to marketed product make it difficult for employers to provide incentives and motivation to researchers in drug discovery, as research scientists usually follow other career options inside or outside their companies or even retire before patients can benefit from their work.

Compared to the long timelines in R&D, the effective patent protection of a marketed drug is relatively short. Patent applications are generally filed during the lead optimization phase of a future potential drug candidate in drug discovery. Given that R&D takes 12–14 years already, the remaining marketable exclusivity is around 8–10 years—a very short time span to generate a return-on-investment (ROI). The U.S. and some countries in Europe have therefore extended the patent term for pharmaceutical products by up to 5 years (patent term extension or supplementary protection certificate). Nevertheless, the trend toward longer development times means that there is an even smaller window to recoup the investments, down from currently about 7–8 years compared to more than 17 years in the early 1960s. As long as the period of exclusive marketability is fixed and constant, given by the duration of patent protection, this phenomenon is characterized by a reverse reciprocal relationship as depicted in Fig. 4.4. Because of the typical bifurcating visualization of this relationship, it has come to be called the 'innovation scissors'.

The average development time for a pharmaceutical product is long mostly because failures become visible only as late as in phases II and III of clinical development. It is therefore the primary objective of the pharmaceutical R&D process to ensure that unsuccessful drug projects "fail early and cheap" rather than "late and expensive." This requires a conscientious management of the entire R&D project portfolio.

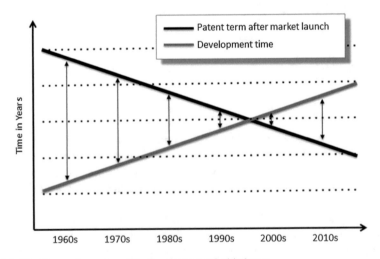

Fig. 4.4 The "innovation scissors" in the pharmaceutical industry

4.2 Phases and Checkpoints in the Pharma R&D Process

The pharmaceutical industry applies highly differentiated phase concepts in R&D. The strictly sequential execution of project phases is sometimes considered impractical, but a given sine-qua-non in medical innovation. Projects are managed operationally following best practice in project management, and are carried out using a gated process with milestones and reviews to ensure candidate quality, proper timing and cost efficiency.

At the multi-project level, the entire pipeline is organized strategically by a portfolio management process in accordance with the prevailing R&D strategy. This scheme appears rigid, and it is so intentionally; nevertheless, we have seen adaptations in how R&D processes are structured. In the past, R&D activities were classified into pre-projects, projects in development phase and drugs in the phase of market introduction. While the management of the latter two could be properly controlled, pre-projects were characterized by a lack of dedicated budget, vague goals and a less structured process.

In view of the high attrition rates in phase II of clinical development, pharmaceutical companies nowadays no longer differentiate and organize their R&D activities in the silos of discovery research, preclinical and clinical development, but aim to integrate early R&D (from discovery to clinical Proof-of-Concept, PoC) and late R&D (phase III or full development).

It is therefore suitable to distinguish three major phases/challenges in the R&D process that need to be addressed vigilantly:

- Concept phase: How to create a good product concept and how to confirm the idea in the human situation?
- Development phase: Given the PoC, how to manage a project to meet the regulatory demands?
- Market introduction: Given the approval, how to transfer R&D results efficiently to operations and the customer?

More specifically, the R&D process usually starts with the idea of a biological target that may produce a clinical effect, if manipulated properly. In vitro and in vivo tests are performed to analyze the potential of the drug target (target identification). Next, the target needs to be verified (target validation), which is usually done by generating a drug-resistant mutant of the target, knockdown or overexpression of the target, or monitoring the known signaling systems downstream of the target. In the following step, a lead compound is identified that has the potential to become a drug candidate (lead finding). Usually, leads are either identified from natural sources, or via HTS or genetic engineering (see also Chap. 3). In the lead optimization phase, the lead compound is optimized in a way that it works more effective in the given in vitro and in vivo test systems. In parallel, early safety and pharmacokinetic analyses are performed in in vitro systems and animals to test the safety of the drug before the human use and to get first result on the absorbance, distribution, metabolism and excretion (ADME) of the lead compound.

The development process starts with the filing of an investigational new drug application (INDA) with the FDA. The results from the preclinical tests, the drug candidate's molecular structure, details to the mode-of-action, potential side-effects, information on the manufacturing of the drug candidate and a clinical trial plan are listed in the INDA. Phase I clinical trials test the drug for the first time in healthy volunteers to assess the safety and the pharmacokinetic of the drug candidate. Phase II trials are performed in 100–500 patients with the disease under investigation to evaluate the optimal dosing and short-term side-effects. In clinical phase III, trials are conducted with up to 10,000 patients to provide statistically relevant data material that allows the assessment of the efficacy/safety potential of the drug candidate. Finally, a new drug application (NDA) or a biological license application (BLA) is submitted to the FDA with the intent to get the drug officially approved. All results and data from the whole preclinical and development program will be evaluated before an approval is given—provided that it passes the scrutiny of the FDA. After the drug is launched, its safety performance continues to be monitored and analyzed to finetune medication recommendations and to investigate in case unexpected serious incidents occur. In addition, phase IV clinical trials may be conducted to analyze specific safety questions or additional effects in patient subgroups. As the drug matures, good life-cycle-management looks for new indications, combinations with other drugs or new formulations to be tested and, if approved, to be commercialized.

Case-in-Point: Risk Assessment of a Drug Candidate: Pantoprazole[1]

Today, Pantoprazole is a marketed drug for the treatment of gastric ulcers and gastro-esophageal reflux disease (GERD), with more than 1 billion patients successfully treated. Pantoprazole irreversibly inhibits the H^+/K^+-ATPase enzyme of the parietal cells in the stomach and hence the secretion of gastric acid is reduced, leading to an efficient pain relief and healing of the ulcerated gastric mucosa.

All registered pharmaceutical medicines have a benefits and risks profile. Administering a new molecular entity (NME) to patients or healthy volunteers is a decision based on the assessment of unfavorable toxicity or side effects compared to the potential beneficial therapeutic effect. The analysis of a NME's preclinical safety profile, especially its carcinogenicity potential, serves the need to assess the risk of a drug candidate before registration.

During the drug development of Pantoprazole, rats and mice were used in carcinogenicity experiments and treated for 2 years with a 100-fold of the intended human dose. While individual rats in this experiment developed benign and malignant tumors of the glandular stomach, mice did not exhibit these tumors, but showed the formation of a few benign hepatocellular liver tumors when treated with the

---

[1]This example was generously provided by Dr. Paul Germann, Head of Preclinical Safety, AbbVie Deutschland.

highest dose. The mutagenicity experiments did not show any genotoxic potential of the drug candidate.

Two questions were of central importance in conducting an adequate risk assessment of these findings:

- How high is the likelihood that humans treated with pantoprazole will develop tumors in the stomach or in the liver?
- How do these data extrapolate and predict their relevance for the intended clinic use?

To evaluate the risk potential of Pantoprazole, the following considerations were important:

- The 2-year treatment of both rats and mice mimic a life-long treatment. Thus, a potential 8-week treatment in humans would correlate to a significantly lower risk in drug exposure.
- Old rats in the sample strain developed chronic nephropathy, leading to a reduced elimination of Pantoprazole through urination and to an increase of the drug in the body.
- No initial genotoxic potential of Pantoprazole was found.
- While liver cell adenomas (benign tumors) are frequent in mice when exposed to a drug for extended periods of time, this tumor is very rare in humans.
- Several studies of various animal species, including rats, mice, hamsters, monkeys, dogs, and pigs, confirmed that rats demonstrate species sensitivity.

Investigations in the mechanism of the tumor formation revealed that as long as Pantoprazole is blocking the acid production in the rat stomach, the increased pH triggers a high production and secretion of the endocrine hormone gastrin. The increased gastrin level leads to stimulation of the ECL cell proliferation and, consequently, to the tumor formation in the rat stomach. Importantly, the proliferation of ECL cells was only observed in rats given very high doses of Pantoprazole. None of the non-rodent species showed this indication. In addition, gastric biopsies taken in human clinical trials and in phase IV post marketing surveillance trials showed no proliferative signals for the ECL cells in humans.

All these factors were considered when assessing the risk potential of Pantoprazole for tumor formation. Based on the broad therapeutic window between the intended treatment duration and dose in humans and the doses which induced tumors in rats and mice, Pantoprazole was considered not to exhibit risk for humans in the intended therapeutic indication. Given the proven species-specific sensitivity (rat, stomach) and low relevance for the human situation (mouse, liver), this compound was approved and registered without any restrictions.

Stage-Gate and Portfolio Management

In the pharmaceutical industry, project management has become very sophisticated and its professional implementation is a key differentiator in project portfolios. Further process-related prerequisites are

- A stage-gate-process with pre-defined milestones at each gate,
- A portfolio analysis process, and
- A portfolio review process.

All research-based pharmaceutical companies establish R&D portfolio management processes to:

- Systematically steer all their innovation projects and related investment decisions,
- Manage the investments in R&D and to ensure a maximum ROI,
- Manage timing and the progress of drug projects,
- Improve the quality of innovation,
- Reduce the overall risk related with R&D projects and
- Control costs of R&D.

The idea of project portfolio management is based on some fundamental considerations:

1. Every drug project can be modeled as a sequential series of phases and gate reviews, at which investment decisions are done based on the results of previous activities and in view of the next steps.
2. Such a model consists of discovery research and drug development the phases of (1) target identification (TI), (2) target validation (TV), (3) lead discovery (LD), (4) lead optimization (LO), (5) preclinical development (PD), (6) phase I, (7) phase II, (8) and phase III of clinical development, as well as the phases for (9) regulatory approval and (10) market authorization (Fig. 4.5).

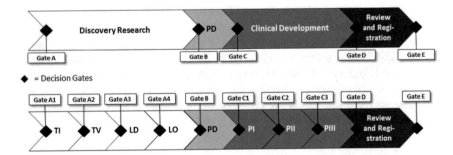

Fig. 4.5 Typical stage-gate-model of pharmaceutical R&D. *PD* Preclinical development, *TI* target identification, *TV* target validation, *LD* lead discovery, *LO* lead optimization, *PI–III* phases I–III

3. A drug project can be stopped in every phase and at each gate of the model (stop option), if the outcome of the project does not fulfill the requirements (as defined in the TPP), or the project does not have the potential to generate value, or if the overall R&D strategy changes. Alternatively, it can follow another path (growth option), if this new path generates less risk or provides more value.
4. An ROI is not obtainable until the new drug is commercialized.
5. The PTRS is defined for each stage of the project: probabilities are available from historical data, benchmarks and expert interviews.

4.3 Portfolio Analysis

The decision whether a project is discontinued or receives more investment, is not taken without analyzing the entire portfolio of R&D projects first. Portfolio analysis starts with the compilation and valuation of all ideas and R&D projects for an overall portfolio overview, but it also includes the portfolio assessment, the definition of a steady-state portfolio, the simulation of the current portfolio in the steady-state-model, the definition of possible portfolio or pipeline gaps in view of the desired project portfolio, the measurement and benchmarking of the performance and productivity (on project and portfolio level) and finally the provision of guidance to line functions.

Traditionally, portfolio assessments rely on qualitative analysis and semi-quantitative metrics, such as the evaluation of the technical feasibility and the market potential of a drug candidate, plotted in a two-dimensional matrix. The portfolio assessment for preclinical research projects can be supported by three methods (see e.g. Betz 2011):

- Criteria-based scoring methods,
- Experts interviews, or
- Open surveys.

An example of a criteria-based scoring method is 3D analysis, which uses pre-defined norms and questions to assess the potential of a drug candidate. The three dimensions of this method refer to (1) the evaluation of the technical feasibility, (2) the maturity and (3) the attractiveness of the drug project. Technical feasibility is a parameter that describes how difficult it is to develop a drug candidate in view of its chemical and biological feasibility, the validation of the drug target, and the patent situation/freedom-to-operate. Parameter maturity describes how close the best of the discovered compounds is to fulfill the criteria of the target product profile. It is related to the questions of efficacy, safety, convenience, formulation, production and availability of biomarkers. Attractiveness of a potential drug candidate is measured by the development money at risk, the competitive situation, the market size and the portfolio fit.

Criteria-based scoring methods are used to provide transparency, and to enable fact-based and rational project decisions. These decisions ultimately affect the expected financial impact of drug projects, the financial consequence of project prioritization, the options for value maximization and correlation with shareholder value creation. Therefore, modern portfolio management increasingly uses quantitative methods to assess the impact of project and portfolio decisions on value creation, and on the risk- and cost-structure of the project portfolio. Based on quantitative analyses, projects and project portfolios are selected that provide the highest overall output possible with the available resources at the lowest risk.

Net-present-value (NPV) is a forward-looking financial indicator for investment decisions (basically, "a dollar today is more valuable than a dollar tomorrow"). Future cash flows need to be discounted towards the present. The later a cash flow is expected to occur, the bigger is its discount factor. Consequently, projects in different phases and with different timelines along the stage-gate-model are valued differently.

In the NPV method, the determinants of value are the expected cash inflows from marketed assets (such as drug revenues and royalties), risks related with R&D, costs of R&D and marketing, timing of R&D and marketing (at least until the date of patent expiration), as well as strategic options arising from technologies and projects. Cash flows that occurred in the past are sunk costs and therefore excluded from the calculation. The risk-adjusted or expected NPV (rNPV or eNPV) method uses this information in combination with decision tree analysis. rNPV is probably the most used method to evaluate R&D projects in the pharmaceutical sector today.

The exact equation and method for NPV calculation is described in more detail by Bode-Greuel and Greuel (2004). Two additional aspects are important in this context: the target product profile (TPP) and the decision tree.

The TPP summarizes all information for a potential drug product that is both approvable and competitive enough to generate sufficient revenues: It is a blueprint of the future product and includes information of the therapeutic area, the patient population, efficacy and safety parameters, formulation and mode of administration (see Table 4.1).

The minimum acceptable TPP defines the criteria that a project needs to achieve at the very least, as they define the regulatory and competitive standards. The desired TPP outlines the best-case development targets of a project. The TPP is the basis for

Table 4.1 Typical template for a target product profile (TPP)

	Minimum acceptable	Desired
Target indication		
Target population		
Primary clinical endpoint[a]		
Secondary clinical endpoint[a]		
Administration route		
Dosage frequency		
Treatment duration		
Onset of action		
Targeted launch		

[a]Items must be described clearly, as for example by giving parameters, amount, benchmarks, and they should be relevant for market success

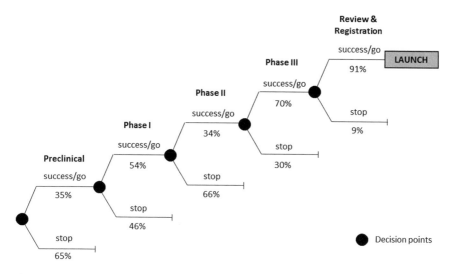

Fig. 4.6 Typical decision tree of a pharmaceutical R&D project with industry-specific probabilities for success (go) and failure (stop)

product development plans and sales forecasts, which in turn are required for rNPV calculations. It supports the definition of target patients and it determines clinical trial endpoints.

A decision tree is a useful tool to illustrate the technical risks and decision options of a project (see Fig. 4.6), including the associated probabilities of all options. It is based on the principles of the stage-gate-model, its phases and decision gates, and it should focus on those activities that are key for the project success and the realization of the goals defined in the TPP. The decision tree includes go and no-go scenarios, and it assigns probabilities to each option, making the likelihood of alternative project scenarios more transparent.

As the decision tree stratifies a project in phases and decision points, investment decisions can be taken with greater clarity on potential risks and consequences. Risks and costs are analyzed per phase and decision point, and the total value gained can be calculated once the project is completed.

Application of a Financial Evaluation of a Project: The Case of Biotech[2]

Most innovative biotech companies do not have the financial resources to take a new molecular or biological entity all the way from research to market introduction. Companies depend on partners, primarily big pharma, to license key products during development.

---

[2]This section was generously provided by Dr. Joachim M. Greuel, CEO Bioscience Valuation BSV.

Both the biotech company (typically the licensor) and the big pharmaceutical firm (typically the licensee) need to agree on a term sheet. Disagreements over the value of a product are frequent, and they result in different ideas concerning financial terms. Solid preparation is important for either party—the biotech or the big pharma company—in persuading the potential partner with fact-based arguments and credible financial analysis, backed by thorough research.

It is common practice to support term sheets with 'comparables', i.e., drug candidates that were similar in overall deal size, upfront payment and royalty rates. Unfortunately, such comparables are rarely a good fit, as reported deal sizes for identical targets, comparable development stages, and geographical coverage vary a lot, thus providing only limited guidance.

It is generally a good idea to invest in a rNPV or eNPV assessment of the project that is to be licensed. To determine a project's value, one needs to compile information on

- The medical need for the intended indication,
- The product's target profile in terms of desired effectiveness, safety and price,
- Already marketed, competing drugs and their market prices,
- Drugs in development targeting the same receptor,
- The product's development plan including timelines and costs,
- Attrition rates in clinical development,
- A company's cost of capital.

The data is used to calculate the product's expected value based on a complex and to-be-defined set of assumptions. The value split between the licensor and the licensee assumes a central role in ensuing discussions. Usually the licensee is exposed to a higher financial risk as he is responsible for late-stage clinical development and marketing, and therefore is accustomed to receiving a higher share of the anticipated product's value.

The use of rNPV in this context has been questioned as it is sometimes believed that input figures lack rigor and may be chosen arbitrarily (especially when a project is early-stage). However, companies experienced with rNPV commit considerable resources to ensuring every single assumption and valuation with hard data and references. For example, development attrition rates can be obtained from several sources and may be adapted to account for a particular disease, the degree of innovation, a drug's mode of action and a molecule's chemical class. Likewise, price assumptions can be derived by comprehensive analysis of market prices for 'standard-of care' treatments, and further supported by high-level pharmacoeconomic assessments. In a market that is increasingly transparent, it is possible to defend every single assumption based on thorough research.

Management should demonstrate a high degree of professionalism by being well prepared for term sheet negotiations. Companies that base their reasoning on comprehensive, evidence-based rNPV assessments combined with reference terms of comparable deals are usually in the better position to define the eventual financial terms. Evidence-based valuations that take advantage of the entire universe of available knowledge take the 'guesswork' out of licensing terms, and ensure each partner a fair share of value.

4.4   Portfolio Management

Portfolio management is a process aimed at maximizing the value of a portfolio of R&D projects through prioritization and subsequent resource allocation (Fig. 4.7). It has its origin in the financial sector and it helps pharmaceutical companies to allocate the enormous resources spent on clinical trials effectively and to the most promising drug projects. Its goal is the identification of projects that both fit to the R&D strategy best and provide the highest value proposition. Thus, modern portfolio management comprises the following elements:

- Financial assessment of projects,
- Analysis of risks related with projects,
- Prioritization of projects,
- Maximization of portfolio value by allocating the available resources to projects according to their priorities,
- The investigation of how changes of the project strategy and of available resources would affect project value, and

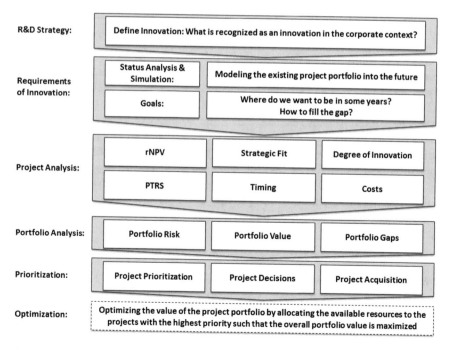

Fig. 4.7 Portfolio management process. *rNPV* risk-adjusted net present value, *PTRS* probability of technical and regulatory success

- The identification of the combination of alternative project strategies that creates the highest value.

Portfolio gaps must be identified and the probabilities of future productivity gaps are determined. Pipeline gaps can be filled either by new in-house projects or by licensing or acquiring external drug projects. The inclusion of new projects need to be considered in context of the overall risk-return structure of the project portfolio; this is not an individual investment decision. The mantra for pharmaceutical companies is to ensure a steady stream of new drug launches, as pipeline and subsequent product flow gaps produce large income losses and result in decline of corporate value, with significant consequences for the organization overall, such as cost cutting, restructuring or divestment.

Project and portfolio assessment, prioritization and subsequent decisions on resource allocation should be done regularly, e.g. as part of annual and bi-annual portfolio review meetings, when a project passes a major decision point in the stage-gate-process, or when the portfolio status and potential is reviewed due to a change management process, such as the merger of two R&D organizations.

The prioritization of individual projects is based on the results of project and portfolio assessments and is typically done in conjunction with a ranking of the projects followed by inclusion and exclusion decisions. Projects with a positive NPV are typically classified as high priority, while projects with a negative NPV are "no-go"-projects. Projects with the highest rank in the prioritization list are in the spotlight of the company. They have the potential to provide the highest value generation, they fit best into the R&D strategy, or they best mitigate overall portfolio risk. Normally, resources are allocated to these projects according to the project budget plan. Projects near the bottom of a prioritization list are considered to have insufficient value potential, are characterized by too many uncertainties too late in their development stage, or do no longer fit the corporate strategy. Typically, projects with low priority are discontinued, put on hold or licensed out. Terminating costly projects sometimes generates value, and in some cases companies have been better off by selling poor-match projects to competitors. Some projects with sub-optimal value in view of the current TPP may be redirected from one therapeutic area to another with a higher market potential, eventually resulting in a better ROI.

It is the absolute goal to develop a portfolio of R&D projects that are cost-efficient as well as risk-efficient, and that provide the highest potential for ROI given available resources; such a project portfolio represents the readiness to assume risk of the investors.

Portfolio Decision-Making at Merck Serono S.A.

The portfolio process at **Merck Serono** covers to all R&D projects and consists of four distinct steps.

In step 1, project claims and strategies are defined. For early stage projects, especially in discovery research, preclinical development and phase I of clinical development, strategic goals are defined using a dynamic option space mapping process that considers several strategic alternatives and respective R&D requirements, and which differs from a typical static TPP. For late stage projects in phases II and III, strategy is defined using a competitive product profile (CPP) including three scenarios with respective probabilities, explicitly the most likely outcome (baseline profile), the downside risks (minimum profile) and the upside potential (optimum profile).

Step 2 is the assessment of R&D risks with six defined risk classes: Conceptual risk, infrastructure risk, compound risk, clinical risk, R&D learning curve risk and operative regulatory risk. The conceptual risk covers all uncertainties of the disease hypothesis, the drug target, and the mode-of-action. Infrastructure risk addresses the questions of freedom-to-operate and technical infrastructure. Compound risk relates to the intrinsic properties of the drug compound. All clinical uncertainties not directly related to the drug compound are summarized under the term clinical risk. R&D learning curve risks address the question of knowledge base and know-how of the organization. By assessing the operative regulatory risk, Merck Serono aims to mitigate all uncertainties arising from the regulatory environment. In sum, more than 500 criteria are used to assess the probabilities and timelines for all R&D phases per project, and to generate input to the decision trees.

In step 3, the commercial potentials for all R&D projects are calculated, either by a qualitative approach to assess the commercial potential (early-stage projects) or by market forecasting and quantitative analysis (late-stage projects).

Step 4 of the process includes a risk return analysis on the project and portfolio levels, the calculation of the portfolio NPV in view of different scenarios, and the analysis of the risk diversification to finally answer the question of what is the acceptable value-to-risk-ratio of the project portfolio that meets best the strategic goals of the company.

4.5 Conclusions

The R&D pipeline is the engine that drives growth in the pharmaceutical industry. Market valuations of pharmaceutical companies are based on prospected new drug approvals and expected new drug revenues. What is feeding these new approvals is a healthy and steady stream of new drug candidates. Given the inherently uncertain nature of research, pharmaceutical companies are introducing not only new technologies to widen the intake of new drug candidates (as described in Chap. 3), but also new management methods and techniques that make drug development more efficient.

These methods center around efficient and effective portfolio management and the use of external resources, which can be accessed either via outsourcing or via collaborations. Typical forms of opening up the innovation process are:

- Venture funding;
- Co-development and joint ventures;

- Strategic alliances;
- In-licensing;
- Out-licensing.

How pharmaceutical companies manage the challenges of in- and outsourcing, as well as the more recent trend towards open innovation, is the topic of the next two chapters.

The Make-or-Buy Challenge: How to In- and Outsource Innovation

5

> *"The process of making a new medicine is a marathon that requires endurance and commitment. We cannot reach our goals without the help of partners from the inside and the outside the company."*
>
> Tadataka Yamada,
> Former Chief Medical & Scientific Officer, Takeda

5.1 The Disaggregation of the Pharma Value Chain

For many decades, pharmaceutical R&D was a fully integrated process in which the pharmaceutical company owned and conducted almost every single task in the value chain. Realizations of economies of scale favored pharmaceutical firms to be vertically integrated, and their R&D processes relatively closed to the outside world.

However, since the 1990s, instead of hiring the best people and doing everything in-house, pharma firms increasingly concentrate on their core advantages and try to involve outside innovators. They followed the trend to actively manage project pipelines by leveraging internal and external assets, as for example by:

- Outsourcing important but non-critical R&D activities that can be done faster, at lower cost, and sometimes at greater quality, by external service providers;
- Joint ventures in R&D, typically focused on co-development or research for a special purpose or therapy area;
- In-licensing of intellectual property (IP) in later R&D stages, often from biotechnology companies;
- Out-licensing of research in the early R&D stages; and
- Spin-offs and divestitures of R&D activities that are either not sufficiently promising or do not fit into the business strategy.

Especially with the advent of open innovation (Chesbrough 2003), a full in-house coverage of all R&D activities seemed no longer desirable or affordable. Here we

© Springer International Publishing AG, part of Springer Nature 2018
O. Gassmann et al., *Leading Pharmaceutical Innovation*,
https://doi.org/10.1007/978-3-319-66833-8_5

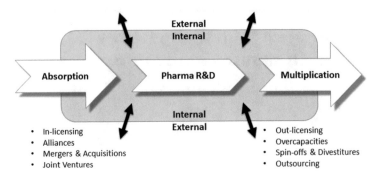

Fig. 5.1 Restructuring pharmaceutical R&D departments and resulting interaction with external partners

focus on the more strategic aspects of pipeline management, namely of how to absorb external and leverage internal assets. We will cover open innovation in more detail in Chap. 6.

Pharmaceutical companies have started to concentrate on core competencies and to focus on certain technology platforms and therapeutic areas to increase the efficiency of their R&D organizations. They are in the process of streamlining R&D activities, reviewing each task whether it should remain inside their own boundaries, which tasks should be absorbed from outside entities, and which activities could be better multiplied by disposing them to external partners (Fig. 5.1). Balancing the right size and structure of the R&D organization has turned out to be one of the primary issues in pharmaceutical R&D management in the last decade. For example, in 2012 **Pfizer** initiated a program to concentrate its R&D activities to five therapeutic areas and closing non-core or less productive R&D areas such as allergies, respiratory diseases, or gene therapy. In addition, Pfizer sold off its animal health and infant nutrition businesses. Today, Pfizer's R&D is focused on cardiovascular diseases, oncology, neuroscience, vaccines, inflammation and immunology; all these therapeutic areas reflect Pfizer's internal strengths, understanding of market trends and external business opportunities, and the aim to provide value to investors.

In modern R&D, all functions are repeatedly analyzed regarding their potential contribution to shareholder value creation. Philosophically, this raises the question about the definition of the boundaries of the corporation. Due to the increasing scale and scope of interactions with outside innovation, every pharmaceutical company's R&D department today is more and more relying on external input from other companies or partner organizations.

This trend of externalizing R&D has led to the creation of several R&D service providers that support pharma firms with technical and scientific services such as R&D contracts, laboratory testing services, technology consulting, industrial design, or even engineering. Firms in high-technology R&D-intensive sectors, including the pharmaceutical industry, have a high propensity to cooperate on innovation projects. The wealth of knowledge generated in new sciences and through new technologies has become too complex for any company to handle alone. Fundamental breakthroughs are increasingly likely to occur not within a single firm's own R&D

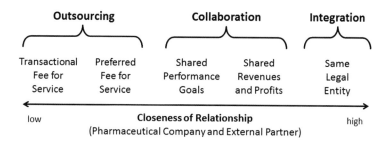

Fig. 5.2 Classification of partnerships in pharmaceutical R&D activities

department but somewhere in an external organization's research lab, perhaps in collaborative arrangements.

Serendipity is still considered a key success factor particularly in the early discovery phases; thus, maintaining a network of outside innovators is very important in pharmaceutical R&D because the likelihood that a company generates all necessary drug candidates in-house is relatively low. Interactions with partners not only reduce the risks of investing in the company's own research infrastructure, but also reduce the risks of inaccessibility to desired drug candidates. Particularly those technology areas that are not yet well covered by pharmaceutical companies are subject to analysis of how they can best be accessed outside one's own corporate boundaries.

The closeness of the relationship between the pharmaceutical company and the external partner serves as a classification criterion of the partners' interaction with the pharmaceutical company (Fig. 5.2). The nature of the interaction can embrace different attributes, features and forms depending on the commitment of resources. In the case of outsourcing, the partner company usually conducts pre-defined tasks in return for a service fee. A collaboration refers to a more closely linked effort between both companies engaged in the joint project; this frequently involves risk, revenue and profit sharing agreements. In the case of an integration, both companies' activities melt together and there is no longer a visible separation between the two entities.

There are multiple reasons why firms establish such alliances. According to Greis et al. (1995), the decision to engage in collaborations is usually based on the trade-off between inter-firm cooperation and vertical integration. This trade-off is seen in the comparison of transaction costs versus development costs. If organizational efficiency gains (due to e.g., shared assets) are greater than production efficiency losses, a firm will choose to collaborate. The general explanation for the overall growth pattern of newly created R&D partnerships is mostly related to motives that 'force' companies to collaborate on R&D. Increased complexity of scientific and technological development, higher uncertainty surrounding R&D, increasing costs of R&D projects, and shortened innovation cycles favor collaboration.

The traditionally integrated structure of pharmaceutical R&D departments is expected to decompose further, and the need to interact with external partners is increasing dramatically. While the complexity of internal pharmaceutical R&D can thus be reduced, the complexity of managing relationships with external partners is rising significantly. Only the pharmaceutical companies that are able to manage their R&D partners optimally will be capable to benefit from the many outside developments and novel opportunities in the pharmaceutical industry.

5.2   Outsourcing of Pharmaceutical R&D

The increasing reliance on outside innovation requires the pharmaceutical company to think and act in a more process-oriented way. Barriers between intra-organizational units and with external partners are expected to lessen. External experts could either be integrated into the innovation process for a limited time or they could just provide some necessary infrastructure and basic services to the pharmaceutical company.

The global drug discovery outsourcing market had a volume of US$15 billion in 2014 and was forecast to reach US$25 billion by 2018 (Kermani 2014). Outsourcing in the pharmaceutical industry is driven by:

- Reduction of over-capacities (as a result of M&A activity)
- Cost cutting or restructuring issues
- Growth aspirations (expertise, resources)
- Reduction of risk and/or proactive risk management
- Corporate governance and/or strategic make-or-by decision

The final decision to outsource R&D functions usually depends on several parameters (see also Festel and Polastro 2002):

- Technological requirements and specifications (technologies and synthesis techniques, available capacities, status of registration)
- Product-specific considerations (quantity, position within the lifecycle, impact on the overall product portfolio)
- Financial aspects (investments, economic feasibility: in-house production vs. outsourced production)
- Taxes (access to medical substances is often used to optimize the tax load)
- Market access (despite new trends, many markets are protected by local governments)
- 'Chemical' tradition of the pharmaceutical company (philosophy, commitment to chemical processes)

If a pharmaceutical company intends to outsource some part of its R&D to an external service provider, three questions are important (see Arthur D. Little and Solvias 2002):

1. What kind of services could potentially be interesting for R&D outsourcing?

2. What are the most effective and efficient interfaces between the pharmaceutical company and the service provider?
3. Which cooperation models are the best basis for managing the outsourcing activities?

The answers to these questions usually dependent upon the characteristics of the pharmaceutical company, its corporate strategic goals and its cost structure. Big pharmaceutical companies, mid-size pharmaceutical companies and start-ups often adopt different outsourcing strategies.

Big pharmaceutical companies are typically involved in outsourcing activities for strategic reasons. In the early phases, this includes process development, scale-up and delivery of first lot sizes for clinical trials. The goal is to circumvent bottlenecks in one's own development process and to manage peak resource shortages.

Mid-size pharmaceutical companies usually concentrate on one or two products emanating from their own R&D pipeline. While outsourcing structures of mid-size pharmaceutical companies are comparable to big pharmaceutical companies, the major difference is the low proportion of outsourced substances of older compounds. Hence, the focus of contract-synthesis lies more on advanced intermediate products rather than on substances.

Start-ups, as the third type of pharmaceutical companies, are typically very much characterized by limited capacities in synthesis development and production. As a result, they almost entirely have to rely on insourcing, and hence represent a huge market for pharmaceutical service providers.

Outsourcing in pharmaceutical R&D is extremely complex, and managing the outsourcing partner can be very cumbersome depending on the field of collaboration. Therefore, many companies try to limit collaboration complexity and use outsourcing only to manage peak resource shortages, despite its potential to help with overall R&D performance improvements.

Outsourcing Strategy: Captive Offshoring at ALTANA Pharma[1]

In 1999, **ALTANA Pharma** set up a manufacturing joint venture in India with Zydus Cadila, an Indian Partner, to access cost effective production capacities via a dedicated site focusing on the complex synthesis steps of active product ingredients (APIs).

Based on the success of the joint venture and due to the need to globalize clinical development activities, a small R&D unit was set up as an add-on to the joint venture to do clinical data management (CDM) and clinical operations supporting global headquarters. This team consisted initially only of a handful of staff, tasked with data entry and query generation. Within 4 years, the Indian team became integrated into

---

[1]This case study was generously provided by Dr. Antal Hajos, CEO Linical Europe.

the headquarter data management unit, which also included a captive data center in Canada.

In 2004, ALTANA Pharma decided to establish a captive Indian R&D center to broaden the service lines in clinical development, but also to utilize the advantages of India as an emerging market with its abundance of scientific talent by establishing discovery services with a focus on medicinal chemistry. The R&D center was hence the start of a new Indian sub-company, spearheading the globalization of R&D into the Indian market. Operationally, the unit reported to the country CEO, with the R&D center formally being part of the global R&D organization.

In a next step, the data unit was removed from the joint venture and integrated into to R&D center. Step by step, over the next 2 years, more tasks and responsibilities were dedicated to this unit, followed by installation of a new global clinical data management system in India to allow compatibility with global clinical research organizations (CROs). These organizational and operational changes were supported by ongoing training of staff, and the building of personal relation within the global R&D organization and in particular to the management at the headquarters.

In 2007, ALTANA Pharma was acquired by the much smaller **Nycomed** for a price tag of 4.6 billion euros. While there were substantial cuts in R&D, the Indian unit with now 130 staff remained entirely untouched, due to its inherent cost advantages. Subsequently, and in view of the cuts in R&D at the German headquarters, the Indian unit was integrated organizationally into the corporate R&D division and took over the responsibility for global clinical data management. The acquisition made Nycomed one of the world's 25 largest pharmaceutical companies.

AJ Biologics: Successful Partnering of a Vaccine Start-Up[2]

Starting a vaccine business requires perhaps more time, expertise and resources than most other businesses. The naturally slow pace and expensive nature of the development, registration and manufacturing of preventive biological products, as well as by the relatively low gross margins and the limited availability of the required expertise, result in a highly concentrated market. Large healthcare multinational companies are maintaining a dominating presence on this healthcare segment, but some start-up companies have mustered the courage and the resources to enter this market space. They are attracted by modest commercialization costs, lasting value of product and manufacturing assets (not entirely related to patents but more importantly to know-how) and of a global vaccine market that is largely demand-driven.

AJ Biologics needed to demonstrate the robustness of its value proposition as it entered partnerships with external service providers. Generally, a biologics start-up becomes a genuinely attractive partner when it demonstrates (1) the ability to access some market segments in selected geographies and (2) the assembly of an experienced team of vaccine professionals.

---

[2]This case study was generously provided by Dr. Pierre A. Morgon, CEO AJ Biologics.

Without having its own R&D and without controlling the manufacturing of its own antigens, AJ Biologics needed to navigate different internal and external stakeholders:

- Politicians and healthcare authorities are valuing self-sufficiency in terms of vaccine supply and the contribution of a high-tech industry to the country's image and economic growth. In consequence, the setup of a vaccine manufacturing capability is a foundational element of the value proposition, reverberating positively on the ability to partner.
- Due to the scientific nature of the vaccine market, the foundational layer is to staff the start-up company with executives having proven technical competences in the field of vaccines. It is also pivotal to select staff that is open-minded, autonomous, entrepreneurial and team-oriented, as most activities are done in collaboration across functional boundaries.
- Successful partnering requires a vision that can be translated into a tangible and achievable reality to ensure that shareholders have realistic expectations and that these are properly managed.

Thus, start-up vaccine companies need solid business intelligence and partnering capabilities, as well as a vision of the target portfolio and a robust command of what technical prowess looks like, in order seek the appropriate partners. The latter will bring technology and some of the investment required for research and development as well as the building of the manufacturing capacity. Some partners will also provide access to the public markets in their country of origin. Due to the long innovation and manufacturing cycles, in terms of the setup of facilities as well as of production lead-times, it is advisable—when affordable—to have multiple partners across diverse projects, to balance the risk and to increase the scope of accessed technologies and markets.

Contract Research Organizations (CROs)

Originally, outsourcing service providers were focused on clinical development, and they were an instrument for big pharma companies to access specialty know-how in preclinical and clinical development. But with the growth of the outsourcing market came a shift in the outsourcing paradigm: from outsourcing 'on-demand' and service provision 'on-request' to strategic outsourcing and preferred partnership models of today. The emerging contract research organizations (CROs) are therefore more diversified and support R&D activities along the whole value chain from research to development, marketing and manufacturing. The CRO market alone was expected to grow to US$35.8 billion by 2020 (PharmaSource 2017).

The major drivers for this development are:

- The need to keep the R&D organizations trim and flexible, and to reduce costs in drug R&D;

- The demand for a global footprint in major countries and emerging markets;
- The need to reduce time-to-market; and
- The realization that only a small number of organizations are able to cover all technologies and tasks for drug R&D by themselves.

Collaborations with service providers have become commonplace in the pharmaceutical industry. Only the biggest pharmaceutical companies may have the resources to achieve a global footprint independently, while biotechnology and specialty pharmaceutical companies need to extend their geographic reach by using outsourcing partners. As early as 2009, more than 60% of the business of most CROs came from small and mid-sized biotechnology companies (Chakma et al. 2009).

Large research-based pharmaceutical companies tend to have numerous drug projects in various R&D phases, and the need to manage this inherent project portfolio risk. They use outsourcing as a strategic tool in all phases of R&D—from drug discovery to clinical development—and therefore prefer so-called preferred service provider and partner align with their internal R&D (full-service outsourcing).

This extensive form of outsourcing is only reasonable if it improves speed and reduces costs (especially for CROs based in emerging markets), and alleviates capacity and capability shortages in drug discovery and development. Typical concerns in such arrangements are complexity and efficiency, IP and royalties, exclusivity and costs. These success factors need to be proactively addressed by both partners and should clarify definitions of efficient and highly standardized processes and contracts. Cooperation agreements often defaults royalties and leaves all intellectual property to the contracting pharmaceutical company. Clear non-disclosure and non-compete agreements (e.g., in specific fields of chemistry) need to be established; as well as comprehensive cost transparency.

No matter which model or degree of outsourcing is applied, the service provider should always be fully integrated into the processes and structures of the pharmaceutical company. Standardization of processes, interfaces, data sheets, incentive systems, etc. facilitates this involvement and makes seamless integration of outside service providers more likely and easier. If those standards are missing and even routine interactions with the service provider are complicated, or if interfaces are difficult to define, the willingness to outsource declines. The establishment of an effective and efficient interface between the pharmaceutical company and the service provider is therefore a critical condition in any outsourcing relationship.

It is important to differentiate between development and production services. Primary development services include process research, process development, and the supply of pre-clinical trial quantities. Primary production services deal with the full-scale supply of intermediates and active pharmaceutical ingredients (APIs) throughout the product lifecycle (Fig. 5.3).

It is to be expected that, in the future, external service providers will be temporarily integrated into internal R&D teams and will then be able to provide more flexibly and timely support for R&D projects. The responsibility for the effective and efficient coordination of all internal and external resources as well as the know-how

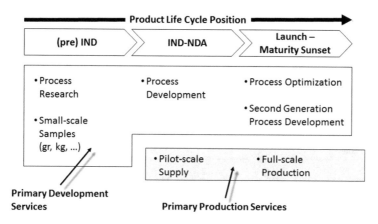

Fig. 5.3 Distinction between development and production services

and expertise transfer will rest with internal project management, which is already regarded as a core competence of the pharmaceutical company.

There will be consequences once this paradigm shift occurs. Barriers between the individual organization units and the various companies will disappear. Pharmaceutical and biotechnology companies will be able to e.g. buy standardized established external services, or new and innovative methods around chemical synthesis along the entire pharmaceutical value chain. Therefore, pharmaceutical R&D will become leaner and will be able to avoid building and maintaining capital-intensive research infrastructure.

Strategic CRO Partners of Pfizer

Pfizer signed strategic contracts with Parexel (U.S.) and Icon (Ireland) in 2011, and added Pharmaceutical Product Development (PPD) as its third preferred strategic partner for clinical research in 2015. Headquartered in the U.S., PPD had more than 50 locations around the world had supported 47 of the global top-50 pharmaceutical and more than 750 biotechnology companies. One of the premier CRO companies, PPD offered both full-service and functional service provider capabilities, including clinical monitoring, data management, biostatistics, statistical programming, pharmacovigilance, medical writing, regulatory strategy and document preparation.

Managing Outsourcing Activities at Solvias

The pharmaceutical service provider **Solvias** was created when Novartis spun-off one of its scientific competence centers in October 1999. In 2016, the company employed over 400 highly qualified pharma experts.

Solvias offered its services mainly in the areas of R&D, production and quality control, focusing on a variety of chemical, physical and biological services—from synthesis to analytics. Customers came from the pharmaceutical, agricultural,

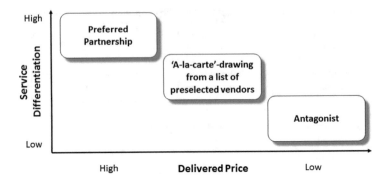

Fig. 5.4 Models of customer-vendor interaction in outsourcing pharmaceutical R&D at Solvias

medical-device, cosmetics, agrochemical and chemical industries, as well as government authorities and research institutes.

Solvias differentiates three cooperation models between the pharmaceutical company and the pharmaceutical service provider, depending on the amount and price of services provided (Fig. 5.4). With the preferred partnership model, the customer (i.e., the pharmaceutical company) enters preferential agreements with a handful of selected, strategic suppliers acting almost as 'facility managers'. However, this model may clash with the customer's desire to keep a certain level of freedom, and to maintain healthy bargaining power vis-à-vis vendors. The second model, the 'a-la-carte'- drawing from a list of pre-selected vendors, is usually applied if no single supplier can provide the breadth of capabilities required by the customer to serve its full spectrum of needs. Overdependence on a single vendor should be avoided in order to spread risks and maintain strategic leverage. The best vendor for each area of activity should be retained by the customer. The antagonist-model provides the customer with the advantage that the vendors (service providers) are systematically put into competition with each other, and hence the best price for the offered service can be secured.

5.3　　Strategic Research Partnerships

The pharmaceutical industry has a long history of collaborating with third parties to access specialty skills, technology and know-how, such as drug target identification, signal transduction pathway know-how, animal modeling, disease expertise, translational medicine and biomarkers. Collaborations between pharmaceutical, with biotechnology companies and with academic institutions have become the norm and are a standard process to increase the knowledge base of pharmaceutical R&D across all technology fields and therapeutic areas (see Table 5.1).

Let us take the example of the research partnership between **Bayer HealthCare**, the German Cancer Research Center (DKFZ) and the National Center for Tumor Diseases (NCT). Initiated in 2009, the academic institutions provided their expertise

Table 5.1 Collaboration deals in the pharmaceutical sector in 2015

Pharma company	Research collaborator	Month in 2015	Scope
Amgen	MD Anderson Center	January	Develop BiTE® therapies for myelodysplastic syndrome
Novartis	Caribou Biosciences	January	Explore CRISPR genome editing technology
AstraZeneca	Harvard Stem Cell Institute	March	Create human beta cells from stem cells for use in HTS screens for new diabetes treatment options
Takeda	CiRA	April	Long-term partnership on induced pluripotent stem cell research
Eli Lilly	BioNTech	May	Develop novel cancer immunotherapies
Bayer	Johns Hopkins Univ.	June	Develop new ophthalmic therapies
J&J	Emulate	June	Use organs-on-chips to improve prediction of human responses in the drug development process
GSK	Francis Crick Ins.	July	Applied research on various diseases
Merck & Co.	MD Anderson Cancer Center	August	Research collaboration in immuno-oncology
Pfizer	Evotec	September	Research collaboration in tissue fibrosis

in cancer research and tumor biology, and Bayer brought in its know-how in drug R&D. The goal of this strategic partnership was the transfer of research results of DKFZ into Bayer's drug discovery and development process, i.e. to improve the pharma company's knowledge base and to fill its oncology pipeline.

This type of collaboration is considered a drug discovery alliance; a long-term (spanning several years) alliance of a research-based pharmaceutical company with a biotechnology company or a research institute to support technology transfer and optionally drug candidate licensing. The advantages of such an alliance are (from the point of view of the pharma partner):

- Bypassing tedious long-lasting licensee-licensor negotiations in subsequent drug licensing;
- Access to internal scientific resources within the parameters of the arrangement, which is an important flexibility given the risks of pharmaceutical R&D;
- The pursuit of R&D novel to the pharmaceutical company, allowing familiarization with a new technology or therapeutic indication without the need to make significant investments on its own; and
- Access to a potential drug candidate, assuming scientific success of the research.

Of course, the academic and biotechnology partners also benefit from such collaborations, primarily by getting access to more technical know-how and resources without the necessity to seek a strategic investor and without the risk of losing control of their own drug candidate.

Another prominent example is **Takeda**'s collaboration with the Center for iPS Cell Research Application (CiRA) at Kyoto University (TCiRA—the TakedaCiRA Joint Program for iPS Cell Applications). iPSCs are induced pluripotent stem cells and have potential clinical applications across a wide range of therapeutic areas including cardiovascular disease, metabolism, neuroscience and cancer immunotherapy. Japan is currently recognized as a leading geographical and regulatory environment to enable stem cell research and to bring research results into the clinic.

Launched in 2015, this collaboration aims at investigating cell therapies with iPSCs. CiRA's director is Shinya Yamanaka, the 2012 Nobel Prize recipient for his work on iPS cells, and CiRA is a world-leading competence center in this field of research. Takeda provides 10 years of funding (¥20 billion, approximately US$177 million), research management expertise, 100 Takeda researchers, access to its compound library and to research facilities at its Shonan Research Center in Japan. T-CiRA has a broad freedom to operate. For example, there is no steering committee of Takeda management to govern T-CiRAS's activities.

T-CiRA is just one of Takeda's major partnerships with academic institutions. For instance, Takeda recently also established a collaboration with TDI in New York, a consortium of three institutions (Cornell University, Rockefeller University, and Memorial Sloan Kettering Cancer Center) to support target research. Takeda also collaborates with Johnson & Johnson by co-investing in the Israeli biotech accelerator FutuRx, with the purpose to fund medical breakthrough innovations that can be spun off in the form of new, independent companies.

Takeda's partnership with CiRA and Bayer's collaboration with DKFZ demonstrate how it is possible to access world-class science by agreeing on a long-term partnership with academic institutions. Conversely, academic institutions access drug discovery expertise, development know-how and marketing power of pharmaceutical companies. Academic institutions such as the University of California in San Francisco (UCSF), Harvard University or the University of Pennsylvania have realized the potential of long-term research alliances and have all formed partnerships with major research-based pharmaceutical companies such as Bayer, Sanofi, Boehringer Ingelheim or Novartis. A special case of academic partnerships is those with university technology transfer groups, e.g. Medical Research Council Technology, Cancer Research Technology, IP Group, and Imperial Innovations.

Importantly, any kind of collaboration, alliance or partnership creates a set of specific advantages and challenges. Among the latter are increased management complexity, coordination costs and risk of IP failure. Normally, academic research—even applied research—is so early-stage with respect to commercialization that substantial additional work and financial investment is required before a return-on-investment can be expected. There is the constant challenge of early result publication, as academia expects findings to be published and enter in the public domain, while companies need to build entry barriers through IP protection. These conflicts are never easy to solve and require tactful maneuvering and mutual understanding.

5.4   In-licensing: Enhancing the Innovation Pipeline

In-licensing is when a pharmaceutical company (the licensee) acquires intellectual property (IP) from another company (the licensor), predominantly a biotech company, to fill a gap in its own development pipeline. Such licenses can be exclusive or non-exclusive. The benefits for the licensor are obvious: The seller receives funds for the traded IP, which is typically used to finance further research. For big pharma, licensing is another major path to accessing new drug candidates, besides in-house R&D, research alliances and M&A. Up to 80% of the R&D pipeline projects of large pharma companies come from external sources—many through in-licensing.

Examples of tradable IP rights include specific biotechnological procedures (such as platform technologies) or compounds. Typical applications of IP are databases or software in which the licensor provides important know-how. Distinctive 'knowledge-service packages' can thus be created and actively marketed. Indeed, patents on novel biotechnological achievements are usually not used to secure knowledge, but to purposefully sell knowledge. Even research results that are not directly related to a specific R&D activity have a certain value and can be marketed. In some cases, complex co-marketing agreements are signed along with the in-licensing deal.

Licensing drug candidates offers important advantages to the licensor, such as strategic pipeline filling, faster time-to-market, higher probability for licensed late-stage projects to reach the market, and an increase in returns-on-investment. Other desirable benefits include:

- Quick expansion of the portfolio of potential drug candidates without the risks and costs involved in a substantial research and development program;
- Better and more flexible utilization of development capacities, which makes the financial risk more calculable.
- Strong complementarity of the in-licensed technologies with those developed in-house (e.g., a business with a promising anti-cancer drug might seek a license of a third party's drug delivery technology to enhance its own product);
- Access to IP rights in platform technologies and software products to enable drug R&D. Pharmaceutical companies often prefer to focus their resources on the later stages of development and commercialization once the potential of a product or technology has been identified, and when the financial rewards are clearer;
- Avoidance of infringement action by third parties. As it is not always possible to work around a patent, negotiating a license and in-licensing it sometimes the preferred way to avoid expensive and potentially disastrous infringement claims.

Pharmaceutical in-licensing has experienced tremendous growth, as companies compete aggressively for promising concepts. Overall efforts in drug licensing increased, deal prices for drug licensing rose and chances of success for deal making decreased. Consequently, companies started to scout for earlier-stage drug candidates, for less commercially attractive late-stage licensing options, and for niche products that include more technical, market and commercial risks. More than 50% of licensing deals are signed for preclinical drug candidates, and early

Table 5.2 Biggest licensing deals in the pharmaceutical sector in 2015 (pot. value in US$ million)

Licensee	Licensor	Disease area	Drug candidate	Phase	Pot. value
Sanofi	Regeneron	Oncology	Immuno-/oncology antibodies	Phase I	2200
BMS	Five Prime Therap.	Oncology	CSFR1 antibody program	Phase I	1700
Sanofi	Lexicon	Endocrine	Sotagliflozin	Phase III	1700
Novartis	Aduro Biotech	Oncology	STING platform	Preclinical	750
BMS	uniQure	Cardiovascular	Gene Therapy Platform	Preclinical	2200
BMS	Bavarian Nordic	Oncology	Prostvac	Phase III	975
Amgen	Xencor	Oncology	Bispecific molecules	Preclinical	1700
Pfizer	Heptares Pharma.	Various	GPCR targets	Preclinical	1890
Pfizer	BioAtla	Oncology	Cond. active biologic antibodies	Preclinical	1000
Biogen	AGTC	Ophthalmology	Gene-basedtherapy	Phase I	1000

Source: Flanagan (2015)

in-licensing is in the strategic focus of pharmaceutical companies. Payments for in-licensing have quadrupled for products in early and late stages of development (see Table 5.2). Average payments have increased, too, with the largest increase for drugs entering the mid-stage trials.

Probably the most prominent example of an in-licensed drug is Lipitor. **Pfizer** marketed Lipitor (initially on behalf of Warner-Lambert before the company was acquired by Pfizer) to compete with drugs such as Zocor (Merck), Pravachol (Bristol-Myers Squibb), and Lipobay (Bayer) for the lucrative cholesterol-lowering drug market. Lipitor was originally in-licensed from Yamanouchi. Pfizer then used its powerful marketing and sales organization to turn this externally sourced 'me-too' drug into one of the most successful blockbusters ever, with total life-time sales of US$141 billion. In 2003, Lipitor became the first pharmaceutical product to top US$10 billion in annual sales. In 2009, just before the end of its patent term, Lipitor generated around US$13 billion in annual sales.

The rise of in-licensing has led to even more complex contractual arrangements. Unidirectional one-time payments have been replaced by sophisticated and timely limited agreements, typically covering different geographical markets. Success premiums, milestone payments, and royalty agreements increase contractual complexity. The extent of the performance-oriented contractual arrangement usually depends on the risk-benefit profile of the collaboration. The earlier the collaboration is entered, the more difficult it is to calculate prospective revenue streams. Many licensors thus reserve the right to market the drug under development by themselves in certain strategically important markets.

In summary, in-licensing of lead substances not discovered internally provides pharmaceutical companies with a vehicle to develop promising drug candidates into blockbusters while it leaves the risk of the initial discovery to an external partner. Therefore, this type of R&D collaboration not only helps the pharmaceutical firm reduce the risks associated with the investments in its own research and discovery infrastructure, but also the risks of not having preferential access to desired substances to fill its own development pipeline. In-licensing always includes the transfer of IP rights and, therefore, links the success of the partners to the success of the joint effort.

5.5   Co-development: Benefiting from Joint Resources

Co-development agreements refer to the mutual development of a drug. This is a type of pharma-pharma or pharma-biotech partnership that is used to complement partner development and marketing capabilities, and to share risks and costs. Pharma-pharma partnerships are also common in the industry and a tool that is often used for co-development of drug candidates to share costs and risks (see Table 5.3 for an overview of recent pharma-pharma partnerships). For example, a biotechnology company that already has a substance in clinical development but does not own a sales force in an important market, teams up with a larger company to jointly develop the last stages of the drug and to sell it mutually after registration. The development is usually conducted by joint teams.

Let us consider the case of **Morphosys** and **Novartis**. In 2004, these two companies started a strategic collaboration to discover and develop antibody-based biopharmaceuticals as therapeutic agents, to address unmet medical needs in a variety of diseases. In 2007, the companies extended the deal to a 10-year agreement, with Morphosys becoming Novartis's main technology collaborator in this area. The financial terms agreed in 2007 included committed payments of US$600 million with the option of an additional payment of US$400 million to Morphosys upon reaching certain additional milestones in the alliance. The price of Morphosys shares increased by 32% after the updated collaboration agreement was announced. Seven years later, as a tangible result from this alliance, the orphan drug candidate Bimagrumab entered the late-stage phases. The muscle-growing treatment is forecast to have a market potential of US$4 billion annually, with further potential in the treatment of COPD, cancer cachexia and sarcopenia.

For many companies, development collaborations are typically driven by a particular project need or by specific market circumstances. They fall into the conducive context of an overall increasing disintegration of the R&D process and the need to develop innovative drugs across various platforms and therapy areas. However, a few companies are now pushing beyond that point, making co-development an integral element of their business model and realizing significant gains in the effectiveness and efficiency of R&D.

Table 5.3 Pharma-pharma partnerships in 2015

Partner companies	Start date	Partnership specifics
Novartis + Amgen	September	Novartis and Amgen plan to co-develop and co-commercialize a BACE inhibitor program in Alzheimer's Disease (AD).
Eli Lilly + Merck & Co.	January	Merck & Co. and Eli Lilly collaborate in oncology clinical trials to evaluate the safety, tolerability and efficacy of KEYTRUDA® (pembrolizumab), Merck's anti-PD-1 therapy, in combination with Lilly compounds in multiple clinical trials.
GSK + Merck & Co.	November	GSK and Merck announced the initiation of a phase I clinical trial designed to evaluate GSK's investigational immunotherapy GSK3174998 as monotherapy and in combination with Merck's antiPD1 therapy, Keytruda (pembrolizumab) in patients with locally advanced, recurrent or metastatic solid tumor(s) that have progressed after standard treatment.
AstraZeneca + Valeant	September	AstraZeneca entered a collaboration agreement with Valeant Pharmaceuticals under which Valeant receives an exclusive license to develop and commercialize Brodalumab, an IL17 receptor monoclonal antibody in development for patients with moderate to severe plaque psoriasis and psoriatic arthritis.
Eli Lilly + AstraZeneca	October	Eli Lilly and AstraZeneca announced an extension to their existing immune-oncology collaboration exploring novel combination therapies for the treatment of patients with solid tumors. Lilly and AstraZeneca will evaluate the safety and efficacy of a range of combinations across the companies' complementary portfolios.
Pfizer + Eli Lilly	March	Pfizer and Eli Lilly are preparing to resume the phase III clinical program for Tanezumab in chronic pain treatment.

For instance, in 2014 **Pfizer** and the German **Merck KGaA** initiated a strategic alliance in the field of oncology/immunology, with the plan to jointly develop and commercialize MSB0010718C, an anti-PD-L1 antibody of Merck's drug pipeline, for the treatment for multiple types of cancer. As per the agreement, the antibody is to be developed as a single agent as well as in various combinations with other Pfizer and Merck drug candidates and approved oncology drugs. Based on an ongoing phase I program, both companies decided to collaborate on up to 20 immunology/oncology clinical development projects, including six clinical trials in phases II and III.

The integration of the operational processes—which work gets done and how decisions are made—are critical success factors for making co-development work. The monetary incentives for the partners to join co-development are usually upfront and milestone payments, as well as royalty payments once the drug has successfully reached the market. Revenue sharing is an important aspect of co-development agreements. Most co-development is done to accelerate development times. The new drug is expected to enter the market sooner than in the case of stand-alone development, and an earlier market introduction translates into higher income even after the partner firm has received the stipulated proportion of the revenues incurred.

Co-development agreements can cover many different areas. A recent co-development agreement between **AstraZeneca** and **Innate Pharma** on Innate's antibody IPH2201 comprises phase II combination clinical trials with an AstraZeneca compound MEDI4736 in solid tumors. Multiple phase II trials are also planned to study IPH2201 both as monotherapy and in combination with already approved cancer drugs. The development of associated biomarkers is also foreseen. AstraZeneca agreed to pay US$100 million before initiation of phase III development, and further payments upon achieving regulatory and sales-related milestones. In this co-development partnership, AstraZeneca is allowed book all sales. Innate is receiving double-digit royalties on net sales, including the right to co-promote the drug in Europe for a 50% profit share.

In general, the success factors for co-development are:

- Development of a business-based co-development strategy based on each partners' strengths;
- Identification of the skill gaps relative to the resources needed for the co-development relationship;
- Assignment of an active sponsor for each co-development relationship;
- Definition of a process and set of criteria for evaluating and selecting development partners;
- Alignment of expectations of the partners, and clarification how the relationship will operationally work in a joint development agreement;
- Determination that each co-development deliverable has a clear, common definition across organizations;
- Establishment of explicit, direct communication linkages between development teams within and across organizations;
- Access to information tools for the development teams to enable secure, real-time information flow between companies including the establishment of processes and organizational elements that facilitate the use of those tools;
- Introduction of regular intervals to measure and assess the progress of each co-development relationship.

In summary, pharmaceutical companies prefer co-development agreements in order to utilize the development and marketing capabilities of a peer. As biotechnology companies increasingly contribute late-stage compounds to their collaborations with established pharmaceutical companies, their enhanced negotiation power enables them to enter co-development agreements as well. The large pharmaceutical company can thus share development risks with partner companies that have already or are about to contribute development to the joint project. Both firms share the benefits of a successful market entry via royalty revenue or profit sharing agreements.

5.6   Out-licensing: Commercializing Internal Research Results

While research alliances, in-licensing and co-development have been common collaboration approaches for most of the research-based pharmaceutical companies, out-licensing has long been considered a difficult task that pharmaceutical companies did not utilize. To some, going through the hassle of selling intermediate research results is not worth the effort, and to others, passing on potentially useful IP to competitors is unwise. However, out-licensing is gaining traction, as recent examples show (see Table 5.4). Even though many pharma companies still lack a consistent out-licensing strategy, it becomes clear that sometimes another company may be better at leveraging a drug than the company that invented it in the first place. As Billy Tauzin, former president and CEO of PhRMA stated, "a medicine that sits on the shelf helps no one."

Out-Licensing: The Hidden Success Factor

One of the most frequently mentioned reasons why pharmaceutical companies have been reluctant for so many years to pursue out-licensing is summarized in Windhover (2003)'s quote that "no one will win any awards within a large drug company for a successful out-licensing deal that generates some upfront and modest expectations for royalties in the distant future." An even bigger obstacle to out-licensing by pharmaceutical companies is the realization that selling the IP rights of a hidden blockbuster to a competitor could be a career-stopper for the executive director authorizing that sale. According to Windhover (2000), most large pharmaceutical companies are uncomfortable about losing control of their drug assets, fearing potential revenue gaps. R&D leaders also worry about selling their drug projects to a competitor.

Another reason is related to information advantage by the seller, and the respective information disadvantage of potential buyers. Why would a pharmaceutical company—which has the necessary infrastructure to develop and market a compound itself—decide to terminate the R&D project if they considered the compound commercially interesting? Research by Kollmer and Dowling (2004) supports this perspective: Fully integrated pharmaceutical companies in their study out-license only non-core assets, i.e., a mismatch of the compound's market outlook and overall R&D strategy. Many potential buyers fail to recognize that some R&D projects are terminated because of reasons not related to the compound itself. Instead of acknowledging the fact that a terminated drug candidate still has a certain medical and financial value, many companies ignore the residual potential of terminated drug projects. The study by Kollmer and Dowling (2004) contradicts this belief and even highlights the potential of out-licensing at large pharmaceutical companies. The results of their research show that the out-licensing activities of fully integrated firms bring comparable compensation to that of not-fully integrated firms, even though the former mostly out-license non-core products. In both cases, licensing seems to be a profitable business. Still, there is much uncertainty to overcome. Research by

Table 5.4 Recent out-licensing deals in the pharmaceutical sector

Licensor	Licensee	Year	Drug candidate	Specifics of the out-licensing deals
Pfizer	Sequella	2015	Sutezolid	Pfizer licensed out the drug candidate Sutezolid after strategic change to focus on vaccines.
Astra Zeneca	Tillotts Pharma	2015	Entocort	Tillotts acquired Entocort, a marketed, locally-acting glucocorticosteroid, indicated for the treatment of inflammatory bowel disease from AstraZeneca.
Astra Zeneca	Valeant Pharm.	2015	Brodalumab	AstraZeneca licensed out psoriasis drug candidate Brodalumab to Valeant after side-effects were identified in phase III trials.
Hamni Pharm.	Johnson & Johnson	2015	HM12525A	Hamni Pharm. licensed out a program of novel diabetes and obesity drugs to J&J.
Novo Nordisk	Bristol-Myers Squibb	2015	Autoimmune research program	BMS acquired a global exclusive license to Novo Nordisk's biologics research program focused on modulating the innate immune system.

Recombinant Capital (2005) on 2623 alliances showed that only one out of eight alliances was an out-licensing deal by a pharmaceutical company (Fig. 5.5).

The Potential of Out-Licensing

As out-licensing utilizes external resources for the further advancement of internally developed substances, out-licensing always promotes the dissemination of technologies and products by integrating a company's intellectual property with complementary assets. Therefore, out-licensing is easier for companies that possess a strong position in a certain technology area but lack the complementary assets necessary to exploit the technology (Fig. 5.6). Successfully executed out-licensing programs provide the pharmaceutical firm with several benefits, such as additional revenue generation, cost and resource effectiveness, or mitigation of R&D related risks. Megantz (2002) states that out-licensing lowers risks because less investment and fewer resources are needed; much of the risk remains largely offloaded onto the partner company, who is now responsible for the further development of the licensed product. If a product fails or a strategy changes, the pharmaceutical company can often walk away without any undesirable follow-up obligation. Termination fees are usually negligible.

The most important reasons for a pharmaceutical company to out-license IP to a third party include:

- Intentions/reasons for internal usage of the technology, compound, or IP are no longer present;

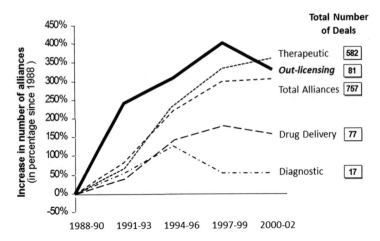

Fig. 5.5   Rise of out-licensing in alliances of top 20 pharma companies

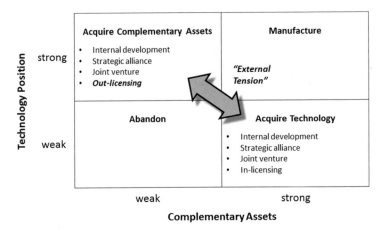

Fig. 5.6   Out-licensing: The neglected strategy to gain complementary assets for the utilization of a company's own technology. Source: Megantz (2002)

- Lack of resources and/or internal expertise for further exploitation;
- The risk profile of the substance and/or compound no longer matches the internal requirements;
- Lack of commercialization potential (e.g., the drug's target market as accessible by the firm is considered insufficient in size to justify further R&D investments);
- Specialization on different product areas or technologies, such as a portfolio restructuring;
- Exploitation of therapy areas other than initially intended therapy areas (e.g., a pharmaceutical company only plans to develop a product or technology in one therapy area, but this product may have applications in other areas; this could

even include areas which go beyond pharmaceuticals, such as cosmetics or plant breeding);
- Low-risk opportunity to move into new, complementary or unknown markets;
- Improvement of the company's revenue stream and/or market penetration by focusing on short-term income;
- Expansion of geographical reach;
- Side benefits, such as increased brand visibility because of advertising by the licensee or the use of improvements developed by the licensee;
- Maximization of the firm's asset utilization and value of the drug portfolio by leveraging internal R&D capacities;
- Gaining advantages in non-core markets by selling non-core technologies;
- Testing a market that may later be exploited by direct investments.

Probably the most important benefit of out-licensing is that the company can increase the utilization rate of its internal research results without using significant additional resources. As soon as the company has made the decision to discontinue internal development of a compound, it should find new avenues to commercialize its intellectual assets at the respective stages of the R&D process. This requires the abandonment of the traditional path of commercialization, and the creation of a new market for the compound (Fig. 5.7). An external partner's R&D resources could be utilized to bring the compound to the market instead of letting it rot away locked up in some proprietary database.

Selling the rights for further development of a compound to an external partner not only helps transfer R&D risks but also allows the generation of royalty revenues in case the licensed compound actually does get launched as a product eventually. According to KPMG, the median royalty rate of licensing deals in the pharmaceutical sector is between 4% and 5% (Courtois et al. 2012).

A critical aspect of any out-licensing agreement is the licensor's ability to retain an interest in the licensed compound's future performance. As the estimation of a drug's commercial and technical potential is very complex and characterized by a high uncertainty, even small errors in the evaluation of this potential can have a big impact on lost revenues. As most pharmaceutical companies are afraid to miss out on the opportunity of participating in a drug's upside potential, they prefer out-licensing deals in which they retain certain re-licensing rights. The most frequently used re-licensing right is the call-back option.

Most pharmaceutical companies today still use out-licensing only as a tactical tool. They only out-license compounds that did not make it into the company's top priority list. However, not only terminated compounds but potentially any compound that can be developed more efficiently by an external partner should be considered for possible out-licensing. A more strategic approach to out-licensing could help improve not only the image of compounds involved, but also the effectiveness of internal R&D.

A more progressive approach to the commercialization of research results could be implemented if the respective out-licensing office received clear profit and loss

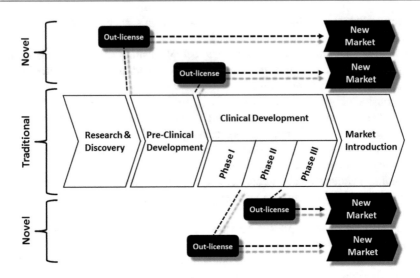

Fig. 5.7 Out-licensing adds novel commercialization opportunities to the traditional innovation process

responsibilities. Out-licensing could be organized as a profit center with a clear mission: To increase the utilization of corporate research.

Managing Out-Licensing at Novartis

At **Novartis**, out-licensing is organizationally embedded in the firm's Business Development & Licensing department (BD&L). This department deploys 80 employees of which three are responsible for out-licensing of substances considered "idle." Novartis usually decides to license out a substance based on the following criteria:

- The substance is no longer strategically relevant;
- The substance does not fulfill the required performance potential;
- New substances arise which cannot be pursued any further within the scope of Novartis and without making significant upfront investments in infrastructure and know-how.

The primary objective of out-licensing is to recoup at least the costs so far incurred in development for the respective substance. In addition, Novartis expects to achieve additional benefits and royalty revenues in case the collaboration partner successfully launches the new substance. For this reason, Novartis always includes revenue participation clauses or re-licensing options in its out-licensing contracts. In order to decide which substances should be out-licensed, Novartis analyses relevant market dynamics and also includes a detailed observation of any competitor activities in the area. Substances that have already dissipated substantial resources are of particular interest for out-licensing as they this could be the last option to generate at least some kind of payback. Therefore, it is not unusual within Novartis that project teams themselves suggest out-licensing a discontinued substance in order to experience at least some satisfactory result for the huge amount of work they put into the substance's research and development.

Novartis sees particularly strong benefits in licensing out to small companies. Small firms are usually very interested in Novartis' substances because their reputation increases significantly if they announce a licensing deal with Novartis. This in turn makes it easier for them to raise funding from other investors.

Out-Licensing in Ten Steps

Novartis uses a standardized out-licensing process (see Fig. 5.8) comprising ten steps. First, a cross-functionally staffed committee shortlists substances with out-licensing potential, considering all globally active development candidates. The head of the Business Development & Licensing department decides on the top candidate. Local teams in the various R&D units worldwide participate in evaluating possible out-licensing of substances targeting their local markets. If the decision committee comes to the conclusion that a substance could offer a strong benefit if out-licensed, the team around the Head of Drug Delivery Licensing & Out-licensing creates a brief product profile of the candidate substance (only around 2–3 pages), aggregating the most important non-confidential product information, mostly about completed development activities and available data on projected potential.

Then, possible licensing partners are identified by screening through a variety of information (publicly as well as internally available) about other bio-pharmaceutical companies and their respective R&D activities. These possible licensing partners also short-listed, evaluated and assessed according to their competencies and capabilities to pursue further development of the drug candidate. This evaluation and assessment already lays great emphasis on—but not exclusively—financial aspects.

Once these possible licensing partners are vetted and considered sufficiently capable, they are being sent the previously prepared product profile with a non-binding inquiry about the out-licensing opportunity. While some candidate substances attract statements of interest quickly, other substances may be available for out-licensing for several years. Eventually, in case of no licensing interest, these unattractive substances are withdrawn and their projects terminated. However, if a possible licensing partner does show interest in a proposed out-license substance, non-disclosure agreements (NDAs) must be signed before proceeding. These NDAs are specific to the substance under consideration and the approached licensees. Only once these agreements are signed, Novartis sends confidential and detailed documentation about the substance. If the targeted partner is still interested after review of this information, the Novartis out-licensing team proceeds to meet with the potential partner in person. Experts from both firms meet and discuss the substance including chances and risks associated with the drug's development. In some cases, Novartis

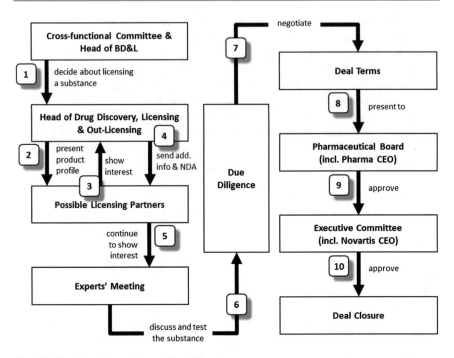

Fig. 5.8 Out-licensing process at Novartis

might also send small probations of the substance (a few milligrams or grams) to the target partner. This allows the partner to proof the substance empirically in in-vitro (i.e., in test tubes) or in-vivo trials (i.e., in living organisms).

Those partners still interested in pursuing a possible licensing of the substance undergo a more detailed on-site due diligence and then, if successful are invited for further negotiations. The typical due diligence takes place at Novartis and lasts about 1 or 2 days. There is no guarantee of exclusivity, and Novartis may negotiate with two or more potential partners simultaneously at this stage still. If Novartis and the potential partner agree on the key terms of the licensing agreement, the Head of Drug Delivery Licensing & Out-licensing presents the case to the pharmaceutical board including the CEO of the pharmaceutical division. If the pharmaceutical board approves the deal, the pharma CEO presents it to the Novartis Executive Committee under the supervision of the global Novartis CEO. Only this committee can ultimately approve the deal. However, the executive committee reserves the right to retrospectively stop or terminate any out-licensing project.

Characteristics of an Out-Licensing Contract

The actual out-licensing contract is prepared by the Head of Drug Delivery Licensing & Out-licensing, a corporate lawyer, an expert from the patent department and a lead scientist. A contract usually covers 50–60 pages. Novartis generally differentiates between two types of contracts: the option-licensing agreement and the regular licensing agreement. The option-licensing agreement gives a partner the option for a product license. In this scenario, the partner can first test whether it can successfully pursue further development of the candidate substance and only at a later point decide about the definite purchase of the license. The regular licensing agreement includes a conventional licensing contract right from the beginning. The contract typically covers upfront payments, milestone payments and royalty revenues. The highest milestone payments occur once the substance finally reaches the market.

The interaction between Novartis and the licensing partner is managed and maintained by a so-called 'key contact person' throughout the collaboration. For some larger deals, Novartis creates a steering committee consisting of BD&L staff rather than an individual key contact person, but with essentially the same management tasks. The key contact person (or steering committee) review semi-annually or annually if the contractual agreements are still met. Typically, Novartis is contractually allowed to withdraw from a licensing deal if certain targets and milestones are not met.

Risk considerations play a critical role in out-licensing at Novartis. While out-licensing represents the risk that an initially internally developed substance might generate significant revenues for an external partner at some point in time, Novartis is also exposed to the risk of not being able to participate in any upside potential for this drug candidate. To manage this risk of a potentially wrong projection of the drug's potential, Novartis usually retains a call-back option in its out-licensing contract. This call-back option can be exercised only at a defined point in time stated in the contract. Different out-licensed substances have different call-back options, depending on product-specific development risks, the substance's market potential, and the performance potential of the licensee. The call-back will usually occur if the risks for further development of the substance are low enough to justify the cost of the call-back and renewed internal development. This point in time is usually between phase II and the end of the development process. In case the point in time for the call-back is not part of the contract, Novartis reserves the right to start re-licensing negotiations depending on the individual situation of the substance's development.

The entire out-licensing process—from initial negotiations of the key terms through to deal closure—usually lasts between 6 and 9 months. Searching for and screening potential partners is not included in this period. Selected Novartis stakeholders are also included throughout the entire out-licensing process, and internal units (especially those affected by the out-licensing deal) are invited to present comments and critique any time.

Structure of an Out-Licensing Collaboration: The Speedel Case

The Speedel case nicely illustrates how out-licensing takes place in practice.

In 1998, researchers at **Novartis** developed Aliskiren, a new substance that was supposed to enter development stages, but Novartis R&D management assessed its potential as not attractive enough and therefore discontinued the project. However, a team of Novartis employees around Dr. Alice Huxley—then Global Project Manager at Novartis—strongly believed in the substance and was interested in pursuing this opportunity further. Huxley and her followers left Novartis to set up their own company called **Speedel**, and eventually proceeded to in-license the substance from Novartis.

Aliskiren (SPP100) is an oral renin inhibitor that demonstrated exciting potential for the treatment of hypertension. In the early 2000s, the market for antihypertensives was approximately US$40 billion or about 40% of the entire cardiovascular market. Hypertension affected more than 135 million people in the developed world. It was (and continues to be) a major cause of strokes, chronic renal disease, congestive heart failure and myocardial infarction. Renin inhibitors, such as Aliskiren, work by regulating the kidney's production of renin. Renin, an enzyme, is associated with the release of a second substance that narrows blood vessels, making it harder for blood to flow through the arteries and raising blood pressure. Aliskiren suppresses the release of renin—and thus keeps blood pressure in a normal range.

Even though Novartis had stopped Aliskiren's development, it was interested in out-licensing the substance to continue its development with a low direct risk exposure to Novartis as the licensing agreement would transfer the development risks to the licensor. In 1999, Novartis out-licensed the substance to Speedel. Speedel was a fresh start-up without an established a track record regarding the successful execution of drug development projects. Thus, Novartis could not be sure whether Speedel would be able to successfully carry on the compound's further development. However, because Speedel's management and core science team had been previously with Novartis, there was considerable trust in Speedel's promise that they would be able to turn the substance into a success.

After deal closure Speedel started clinical phases I and II for Aliskiren. In total, Speedel conducted 18 clinical studies with about 500 patients and healthy volunteers. In addition to pilot studies in chronic renal failure and heart failure, Speedel ran a 4-week, 220-patient phase II study that compared the compound's performance to Merck's Losartan (Cozaar); the two substances showed similar blood-pressure lowering effects. Speedel was also the first company to establish clinical proof of concept in phase II and to have developed and patented a commercially viable manufacturing process for a renin inhibitor; this had been a research focus in the industry for over 20 years.

Throughout the duration of the collaboration, Novartis retained a call-back option to license the substance back at any stage of development. As Novartis was increasingly impressed with the development results, the company exercised this call-back option in September 2002 (see Fig. 5.9). In November 2003, Novartis announced that its was moving Aliskiren into clinical phase III trials, which were initiated in March 2004. This left Speedel with significant milestone payments due to the successful development in phase I and II. (The amount of the milestone payments that Speedel received from Novartis in July 2004 was not disclosed, but they included a cash element and an equity participation.) In January 2005, Novartis announced positive data from the phase III study of Aliskiren as a monotherapy for

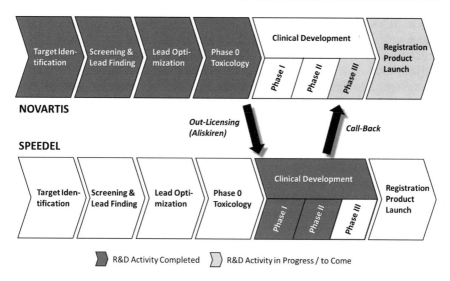

Fig. 5.9 Out-licensing collaboration between Novartis and Speedel

the treatment of hypertension, and positive phase II data from a combination study of Aliskiren with Diovan, another anti-hypertensive compound and a leading drug in the Novartis portfolio. Novartis filed for regulatory approval in 2006, receiving market approval for Aliskiren in the U.S. in 2007.

Speedel benefitted from this success and became a publicly traded company at the Swiss Stock Exchange in 2005, employing around 75 scientists and support staff. In 2008 Novartis, owning already 9.7% of Speedel stock, offered to buy out the five largest private stock holders, including Dr. Huxley, Speedel's founder, for nearly US$900 million. Since September 2008 Novartis holds 99.8% of Speedel stock, with Speedel now operating as a Novartis subsidiary.

In summary, out-licensing enabled Novartis to increase its 'shots on goal' without simply enlarging its already significant in-house R&D budget. The success of the collaboration with Speedel demonstrates how the continuing disintegration of the pharmaceutical value chain creates new partnership models that allow pharmaceutical companies to leverage not only their own internal core competencies, but also the strengths of their partners.

5.7   How to Commercialize a Breakthrough Technology

As an illustration of a partnership between two companies to exploit the full potential of a new technology, the following case describes the commercialization of InnoGel, a breakthrough soft capsule technology, and covers in detail the structure of the commercialization process.[3] The commercialization of InnoGel was performed through a joint collaboration between **NovoGEL** and **InterPharmaLink**.

The InnoGel technology was developed by NovoGEL, a spin-off company, which holds all rights to the InnoGel technology and acts as a licensor towards interested parties. InterPharmaLink is a specialized healthcare management consulting firm that has the exclusive and worldwide mandate to commercialize the InnoGel technology. InterPharmaLink helps clients optimize key resources along the value chain and create value added through product portfolio optimization, supply chain optimization and business development.

The Soft Capsule Market

InnoGel was developed and commercialized in the mid-2000s. At that time, about 100 billion soft capsules were produced throughout the world per year. Virtually all of them were manufactured by contract manufacturers and accounted for a market volume of about US$1.5 billion. Soft capsules were mainly applied in the industries of pharmaceuticals, health and nutrition, and cosmetics. The soft capsule market was dominated by two contract manufacturers: R.P. Scherer, the originator of the rotary die soft capsule manufacturing process and Banner Pharmacaps.

Soft capsules were a means of packaging whereby the capsule content was encapsulated by a shell. The state of the art technology for the soft capsule shell was gelatin. The development of a non-animal alternative to gelatin had been a top priority in the soft capsule industry for many years. Several attempts to replace gelatin by non-animal material had been made previously, but none of them had been overly successful. The InnoGel technology was a starch based gelatin replacement technology that had the potential to revolutionize the soft capsule market as it offered new unique material properties and significant cost advantages versus gelatin.

The InnoGel Technology

The InnoGel technology was based on starch gel, an ordered material with a partly crystalline network structure. The network density was adjustable by varying the composition and the physical treatment of the starch gel. It is this unique feature of the network structure that led experts to believe that the InnoGel technology would be able to produce superior soft capsules for health and nutritional as well as for pharmaceutical products. These improvements manifested themselves in various strategic and financial benefits:

- Additional market potential due to suitability for vegetarians and cultures averse to animal sources (bovine or porcine);

---

[3]This case study was generously provided by Dr. Marc Müller, InterPharmaLink.

- Increased value proposition vs. customers due to superior capsule properties and suitability as a lifecycle management tool for pharmaceutical products;
- Improved safety for end-users and reduced risk profile for customers due to elimination of BSE risks;
- Significantly lower raw material cost due to lower price of starch versus gelatin;
- Significantly lower process cost due to higher yield, less rejections, shorter process time, lower energy consumption and easier handling;
- Significantly lower packaging and logistic cost due to superior capsule properties.

The Commercialization Process

The aspiration of the commercialization was to capture a maximum share of the value creation potential of the InnoGel technology. The commercialization of the InnoGel technology was to be performed along a clearly defined process following three major process steps:
   Step 1: Define the basic conditions of the commercialization;
   Step 2: Develop the strategy of the commercialization;
   Step 3: Manage the out-licensing process.

Step 1: Define the Basic Conditions of the Commercialization
At the beginning of the collaboration, NovoGEL and InterPharmaLink agreed on the important conditions of the commercialization of the InnoGel technology with regard to confidentiality, timing and financials:

- Only non-confidential information was to be disclosed to interested parties as the patent application ad been only published 4 months after project start.
- Only very limited financial and personal resources were to be applied to the commercialization of InnoGel, as NovoGEL was a start-up company with limited financial capabilities.
- The first license was to be granted within 12 months after project start in order to ensure the financial stability and future projects of NovoGEL.

   In addition, both parties agreed on a collaboration agreement with clear incentives for both parties to fully exploit the value creation potential of the InnoGel technology and to perform the commercialization process in best time and at lowest cost.

Step 2: Define the Strategy of the Commercialization
The goal of the strategic planning of commercialization was to fully exploit the value creation potential of the InnoGel technology while considering the given conditions (i.e., tight schedule of commercialization and limited applicable resources). A successful commercialization strategy is usually based on a thorough analysis of the strategic landscape and addresses the strategic questions of where, how and when to compete.

The major soft capsule manufacturers were identified as the primary target group for commercialization. These contract manufacturers possessed an extensive know-how and long-term experience in soft capsule manufacturing and had virtually all necessary equipment available to perform the technical evaluation, as well as the subsequent application development in best time. The most important selling arguments for contract manufacturers were cost saving potentials on the raw materials, as well as strategic aspects regarding BSE risks and issues. The soft capsule manufacturing process was mainly operated by contract manufacturers serving companies in the pharmaceutical, health and nutrition or cosmetics industry.

A secondary target group were companies with considerable capsule volumes wanting to reduce or eliminate their dependency on contract manufacturers by re-integrating the soft capsule manufacturing into their in-house production facilities.

The licensing-out of the InnoGel technology had been identified as the most promising approach for commercialization. As the InnoGel technology was still in the development stages and the future licensee would have to invest in ongoing development, a combined technical evaluation and option for licensing agreement was considered the most appropriate alternative. In order to ensure maximum know-how protection for licensee and licensor and preserve the exclusivity of the know-how, the out-licensing was done as a structured bidding process among interested parties.

The timeline of commercialization was set at the beginning of the project, meaning the first license was to be granted within 12 months after project start. This early decision was mainly driven by the aspirations of the NovoGEL business plan.

Without these given constraints, a detailed assessment of the impact of the timeline on the exploitation of the value creation potential would have been the appropriate way to proceed. Under immediate commercialization the value creation potential for NovoGEL was clearly limited, as the InnoGel technology was an early development stage project with unproven realization guarantee. On the other hand, the risks for NovoGEL were clearly limited as well because the complete application development would have to be performed by the future licensee.

Step 3: Manage the Out-Licensing Process

In order to further develop and commercialize the InnoGel technology in a way that would give maximum protection of know-how to licensee and licensor, the out-licensing of InnoGel was done as a structured bidding process among the top global soft capsule manufacturers. The exclusive rights granted consisted of the rights to perform an exclusive technical evaluation coupled with an option for subsequent licensing. In order to preserve the exclusivity of the know-how, the commercialization process incorporated the following principles:

- The exclusive technical evaluation was to be linked with the option for licensing. Therefore, the future licensee would have to sign a combined technical evaluation and option for licensing agreement prior to starting the exclusive technical evaluation phase.

- The detailed know-how for the exclusive technical evaluation (e.g., the recipe of the starch gel), as well as the process parameters would only be provided to the party signing the combined technical evaluation and option for licensing agreement.

The technical evaluation and option for licensing agreement incorporated the conditions of the exclusive technical evaluation phase as well as the conditions of a subsequent licensing agreement. Interested parties were offered the following conditions for an exclusive technical evaluation phase and subsequent licensing:

- The exclusive technical evaluation phase is limited to a 3-month period and based on a detailed work plan describing the planned trials as well as the timelines to be met for the results. Both parties agree to 'target outcome results'.
- The technical evaluation is to be conducted by an independent committee on the basis of these predetermined 'target outcome results'. Subsequently, the evaluating party are to decide whether it wants to exercise its option for licensing.
- All intellectual property rights and patents as well as results derived from activities during the technical evaluation phase are to belong to NovoGEL. If the licensing option is exercised, the rights and results would be part of a subsequent license.

The decision with whom to enter concrete negotiations to sign the combined technical evaluation and option for licensing agreement was based on a set of qualitative and quantitative aspects; particularly the following:

- What timeline would be agreed upon and what trials were performed during the technical evaluation period?
- What technical evaluation fee was offered when signing the combined technical evaluation and option for licensing agreement?
- What down payment and/or exit fee was offered? A down payment was payable if the technical evaluation is successful and the evaluating party wanted to exercise the option for licensing. An exit fee was payable if the technical evaluation was successful and the evaluating party did not want to exercise the option for licensing.
- What first and second milestone payments were offered? A first milestone payment was payable when the first health and nutritional product was on the market. A second milestone payment was payable when the first pharmaceutical product achieves regulatory approval.
- What royalties were offered? Royalties were payable based on the number of InnoGel capsules sold.

The decision with whom to enter negotiations for signing the combined technical evaluation, and the option for the licensing agreement, was based on the attractiveness of the offers submitted by interested parties.

Lessons Learned

Lessons learned from the successful commercialization of InnoGel can be summarized in the following three recommendations:

- Define the basic conditions and requirements of the commercialization with regard to timing and financials and set-up a detailed work plan;
- Analyze the market and develop a commercialization strategy that provides clear answers to the questions of where, how and when to compete and respect own resources;
- Set up a well-structured out-licensing process that ensures the capture of a maximum share of the value of the technology for the licensor while preserving the exclusivity of know-how to the licensee. Conduct open and fair negotiations.

5.8   Conclusions

The disintegration of the pharmaceutical value chain continues, and has literally created new industries with new service providers and new benefits for all pharmaceutical firms involved. Even in the relatively conventional domain of collaborations and partnerships, out- and in-sourcing, as well as out- and in-licensing, require a well-crafted strategy and should no longer be left to opportunity and chance. Only companies that master the bread-and-butter of the complex multi-faceted pharmaceutical innovation process will be able to succeed in the even more challenging context of open innovation and virtual partnerships. In other words: a pure shopping mentality is not fully using the potential of open innovation. In order to absorb, digest and commercialize knowledge from the outside, strong internal capabilities in R&D are required, as well as expertise in alliance management.

The Open Innovation Challenge: How to Partner for Innovation

6

> *"We are passionate about working openly with the best scientists around the world to understand key pathways and mechanisms that can help transform great science into great medicines."*
>
> Mene Pangalos,
> EVP, IMED Biotech Unit, AstraZeneca

6.1 From Closed to Open Pharma

Traditionally, R&D in the pharmaceutical industry relied on its own corporate scientists, proprietary technologies and internal know-how. Target identification and starting points for new drug projects came from in-house experts or were internalized through patent monitoring and surveillance of competitors. Target validation, hit finding, lead discovery, lead optimization and preclinical development were all done inside the company. In view of the stringent regulatory environment, the high technical risks and the enormous costs associated with drug R&D, pharmaceutical companies typically used stage-gate-models to select the most promising drug candidates. Resources were allocated to the projects with the highest probability of success, while low priority projects were terminated. Consequently, the number of drug projects decrease along the R&D funnel from drug discovery to development. Clinical development was managed by internal staff and executed with support of clinical research organizations. Pipeline gaps were filled by licensing in drug candidates. Risks and costs associated with drug projects were shared with co-development partners.

We have already talked about the many challenges that pharmaceutical companies face in managing the process of discovering and developing new drugs: empty development pipelines, declining R&D efficiency, cumulating pressure from payers for lower drug costs, and increasing R&D complexity. These challenges are not unique to the pharmaceutical industry, but given the unusually long lead times in pharma innovation, they pose exceptionally great risks to the financial, strategic and technological foundations of pharma companies. The advent

© Springer International Publishing AG, part of Springer Nature 2018   111
O. Gassmann et al., *Leading Pharmaceutical Innovation*,
https://doi.org/10.1007/978-3-319-66833-8_6

of 'open innovation' (Chesbrough 2003) in the early 2000s—based on the idea that innovation is more about leveraging a network of innovation contributors than owning and controlling every single innovation task internally—created therefore much excitement and optimism that this new management approach would help reinvigorate innovation also in the pharmaceutical industry.

These expectations have been mostly fulfilled. On the one hand, open innovation is flexible enough as a concept to encompass both incremental and radically new approaches to collaborative R&D and innovation, making it easy for pharma managers to experiment with new organization forms. The trend towards open innovation was also facilitated by the increasing IT savviness of researchers, patients, and pharma partners. On the other hand, open innovation also rattles some of the most espoused value systems within pharma, such as the ownership and protection of intellectual property (IP) or management of drug uncertainty, risk and safety. Nevertheless, pharmaceutical companies nowadays experiment with various initiatives and tools that are based on, or inspired by, open innovation:

- Venture funding of early research and potentially breakthrough innovations;
- Create growth options by accessing external drug projects by licensing or M&As;
- Partner with world-class academic institutions to access specialty know-how;
- Establish innovation centers and innovation camps to push creativity;
- Virtualize R&D and outsource R&D activities to preferred providers to increase R&D efficiency;
- Disclose problems, transfer knowledge and share IP with partners and the crowd to reduce time-to-market; and
- Manage the R&D pipeline strategically to leverage the output/input ratio of R&D investments.

More specifically, the business models of major pharmaceutical companies increasingly include external R&D elements that range from the acquisition of drug projects, traditional collaborations with academic institutions, multi-year research partnerships, crowdsourcing, centers for R&D excellence, open source innovation, virtual R&D and public private partnerships (PPPs).

6.2   Dealing with Uncertainty in Virtual Organizations

People at the Center of Dealing with New Forms of Innovation

Science is—somewhat contrary to public opinion—preoccupied more with the *lack* of knowledge than with knowledge itself. This is especially important in science-driven industries such as the pharmaceutical sector. Chapter 3 introduced various new techniques and technologies developed to close knowledge gaps and acquire and process data quickly. Despite the impressive advances made in new technology, it is the skills, know-how and experience of scientists and research managers that make the difference.

Therefore, pharma management has always taken a strategic approach to managing researchers and R&D supervisors. The management of human resources (HR) in pharmaceuticals has traditionally focused on developing specialists and reducing the (unwanted) flow of information beyond predetermined boundaries. More recently, HR management has started to include, among others, personnel assessment and training, knowledge management, and strategic leadership development. These aims necessitate a more open-minded approach to managing people.

A successful HR management approach creates an environment where every employee in R&D asks himself what his or her contribution could be across all levels of the R&D process. This approach may require abandoning the structured linear process and move towards more group- and team-oriented R&D work. Feedback loops (i.e. from the clinical trials back to basic research and the screening stages) need to be established. Post-project reviews identify not only technical but also managerial areas of improvement. The overall goal of R&D should not be to be innovative but to generate products that are successful on the market.

HR has an obligation to develop future project managers and business leaders. HR should act as a consultant to department heads and group managers how to best develop and promote talented subordinates. Exit interviews still focus too much on the past and too little on improvements for future positions and future successors. Where HR faces resistance from conservative functional managers, it should insist on the bigger picture of developing motivated and educated individuals rather than stagnating technical functionaries. Strategic human resource management in pharmaceuticals addresses these issues upfront, and develops a career roadmap for talented individuals who can grow through R&D projects, business development, as well as marketing and sales experience, in order to become the future R&D leaders that pharmaceutical companies will need 10–15 years later.

The end of the cold war has ushered in a paradigm shift in science. Fundamental industrial research funding drizzled out, and companies needed to find new sources to support expensive R&D programs. At the same time, new entrepreneurial opportunities allowed young researchers to start commercializing their technologies themselves. Science became more open partly out of necessity (research labs reaching out for collaboration partners to sustain their innovation funnels) or out of opportunity (graduate students turned entrepreneurs seeking complementary technologies to start new businesses). While the pharmaceutical industry has followed or even developed this trend in outsourcing some of their basic research to biotechnology companies, they have often failed to rethink their internal R&D organizations to reflect the new conditions.

Small, multi-disciplinary teams have been at the core of the success of biotechnology start-ups or the introduction of new technologies. Large research and development departments are often good at administrating long-term research efforts, but incapable of flexible decision-making and absorption and dissemination of innovative ideas. Job rotation and training programs educate the individual researcher about other activities in the company, and allow him or her to make better and faster decisions: i.e., making decisions where the action is. For example, Bayer is known for their job rotation practice, where every researcher is only allowed to stay for a maximum of 5 years in the same position. Japanese pharmaceutical companies are also applying sophisticated job rotation programs. However, in many pharmaceutical companies, most of the researchers only have contact with the HR department only once: when they are hired.

Fostering creativity is another important aspect in managing human resources. Scientists spend only an estimated 11% of their time on research, and only 2% on new research. It is a long-established fact that interdisciplinary teams are better at innovation: Chemists, microbiologists, medical doctors, and marketing experts should collaborate in new projects to learn from each other about the nature of their businesses. Furthermore, creativity is generally greater in smaller research teams. This should be kept in mind given mega-mergers and building ever larger companies.

"Virtual" R&D as a Form of Flexible Engagement

In the context of organizational structure and collaboration, the term "virtual" is associated with these highly-related notions: changing and transparent organizational boundaries, supported by information and communication technologies (ICT), dispersed in space, with changing degree of collaboration intensity (Maznevski and Chudoba 2000). Virtual R&D is facilitated by shared ICT networks (to expedite information and data exchange), the standardization of interfaces, and the high quality of the work carried out by the different members of the network (Gassmann and von Zedtwitz 2003a).

Information in the form of knowledge about technologies and products, customer feedback, product tests, research results, and markets, all must be collected, analyzed, and transferred. This information must be available in a codified, quantified or explicit form (Nonaka and Takeuchi 1995). Implicit or tacit knowledge, which is difficult to articulate because of its ambiguity and context-relatedness, is much harder to transfer. Systemic innovation with its late 'freezing points' requires the interaction and integration of different knowledge sources and presents a great challenge for multi-site R&D.

Therefore, R&D can be 'virtualized' only if innovation is autonomous, if there are few interdependencies between parallel work tasks. Systemic innovations (i.e., R&D that involves the realization and adaptation of complementing technologies with complex interfaces) are more effectively done in large companies with central R&D. If an autonomous innovation is carried out by a centralized organization, the company is usually outrun by small firms or large decentralized companies.

Virtual R&D has thus grown in importance, especially in industries characterized by rapidly advancing technologies, relative scarcity of technical talent, and the presence of more and more codified information in the innovation process: mostly in the electronics, information technology, software, pharmaceutical and chemical industries. New expertise and basic research are brought into research organizations through collaborations and interactions with universities and small biotechnology companies. In addition, more emphasis is placed on providing novel tools for research such as high-throughput screening, genomics and combinatorial chemistry.

The goals of virtual R&D partnership are a higher flexibility to choose the best R&D service providers, spreading of risk and costs, and economies of scale and scope. Companies try to reduce their fixed costs through transfer of less intensively utilized services to learning partner companies. This has two major advantages, namely

1. A reduction of internal complexity due to concentration, and
2. A reduction of external interfaces, since the task is now shared with the partner companies.

In the following examples, we describe how pharmaceutical firms have chosen to leverage virtual organizations to address some of the challenges emerging from increased operational and strategic uncertainty, and the need to tackle systemic complexity in multi-partner collaborations. These solutions can be either intra- or interorganizational; we begin with the case of the internal virtual R&D project management pool at Roche.

Managing Virtual Project Management Pools at Roche

The systematic promotion of human resource development strategies can be backed up by a specialized project management department as applied by Roche (Fig. 6.1). Roche's Pharma Division established a department called International Project Management, tasked to coordinate a resource pool of about 50 project managers for R&D projects worldwide. This department consumes about 30% of the pharmaceutical R&D expenditures and has high strategic importance to the innovative potential of this business.

Every project manager belongs to this geographically decentralized department, receiving specialized training to lead global projects. The director of this virtual resource pool assigns managers to projects as part of a global program to ensure standards in quality and project procedures. Upon completion of the project, the project manager returns to the resource pool. As there are more projects in the pipeline than managers available, they are immediately reassigned to a new project.

Since this new department reports directly to the board, the internal position of R&D project managers is improved. The establishment of a project manager pool is a clear signal for empowering one of the scarcest competitive resources. Roche thus retains much of the valuable procedural know-how of conducting and leading international projects. Not only is this done on a project level, but it is also placed in a position where it can be reapplied when needed. Project offices are especially valuable when projects are very long—sometimes up to 15 years in the pharmaceutical industry. No individual is able to carry out more than 2 or 3 projects. In a project office, one can learn from dozens of previous and ongoing projects.

The director of the International Project Management Department is also a member of the International Project Committee, which decides over more than 60 global R&D projects at Roche Pharma. The director assumes a role as interpreter or liaison between project managers and top management, thus representing the interests of international project management at Roche. This virtual pool promotes

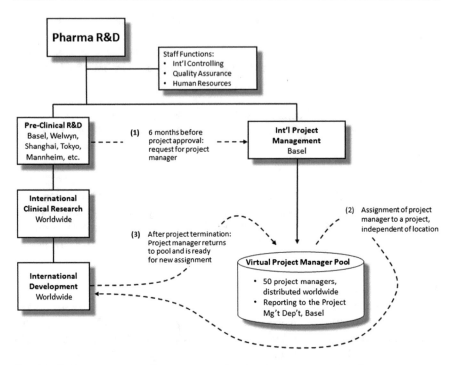

Fig. 6.1 Project management department as a virtual project management pool

the project management idea: experienced project managers are dispersed around the world and move from project to project, no matter where the next project will be conducted.

Virtual Pharmaceutical R&D Organizations

A virtual pharmaceutical R&D organization is an enterprise with only a small number of internal staff that uses external resources to realize its R&D projects. It obtains ideas and drug concepts from both internal and external sources, but it executes its R&D projects primarily with external resources that are sourced globally and on demand with the benefit to save costs, leverage the best value out of the project and progress the projects in faster time. Collaboration partners help with drug commercialization. Such virtual companies are also defined as alliance networks, or as companies that outsources everything.

At the core, the operating model of a virtual organization is to bundle external service providers to progress drug projects along the value chain. The advantages of virtual pharmaceutical companies are:

- Instant access to new technologies,
- Reduced capital requirements,
- Mitigated financial risks,
- Reduced overhead costs,
- Limited infrastructure costs,
- Flexibility in selecting the most suitable contractors and experts,
- Simple governance structure,
- Lack of bureaucracy,
- Quick decision-making, and
- Reduced time-to-market.

Companies that use the virtual R&D model do not need to build up a permanent R&D organization (headcount and infrastructure) and can access flexibly external resources in close alignment with pipeline needs. Any service provider available on the global market is a potential collaboration partner in drug R&D. The virtual company is thus a project-oriented organization enabling each project-team to act more independently, more flexibly and faster in decision-making. The core of a virtual R&D organization is a highly flexible expert team of senior scientists and industry experts.

To operate a virtual R&D organization successfully, one should have:

- Strong industrial and academic networks,
- Professional project management skills,
- Outperforming licensing skills,
- Highest expertise in IP management, and
- Professional expertise in financial valuation of drug projects.

Examples of External Virtual R&D Organizations

In 1996, Roche established the London-based **Protodigm**, one of the first 'virtual drug development companies' in the pharmaceutical industry (Hofmann 1997), with the aim to increase capital-efficiency and cost-effectiveness, and to reduce time-to-market of drug development. This was part of Roche's redesign of the R&D process to involve more specialized partners and suppliers, or to set-up focused R&D service providers itself. Examples of spin-offs of corporate R&D at Roche include Actelion and BioXell. Protodigm was renamed **Fulcrum** in 2001, expanding to offices in Japan and the U.S.

Ten Protodigm employees simultaneously oversaw up to three future drugs in various stages of pharmaceutical development. For instance, if a certain molecule (a prospective new medical substance) was discovered in a university laboratory, rather than acquiring the compound and the underlying team outright, Fulcrum would facilitate further research by guiding the R&D process, contacting specialized companies to test the substance, coordinate the first clinical trials, and contract out production, second-stage clinical development, manufacturability tests, drug registration, and even sales. Up to 15 specialized service providers would be brought in for e.g. regulatory and licensing, quality assurance, market research and marketing.

Essentially, Fulcrum manages a team of external contract research organizations (CROs) and contract manufacturing organizations (CMOs). All of these various subcontractors were then welded into a virtual boundary-spanning innovation team. Since Fulcrum chose the most qualified subcontractor for each stage of R&D, Roche expected a reduction in R&D costs of up to 40% without jeopardizing the already tight development schedule.

Roche was not the only pharmaceutical company experimenting with virtual R&D: Merck was said to have saved US$170 million with this type of outsourcing in 1996 alone. **Eli Lilly** has also been very active in building a new R&D model to address the efficiency challenge of research-based pharmaceutical companies. Supplementary to its crowdsourcing initiatives Innocentive and YourEncore, Eli Lilly started an experiment in 2002 that uses the strengths of a biotechnology start-up company in flexibly managing R&D projects combined with its own expertise and power to develop new drugs globally.

Adding to its in-house R&D organization, Eli Lilly founded **Chorus**, a small and independent entity that focuses on managing drug candidates from candidate selection to Proof-of-Concept (PoC) at low costs and provide the candidates for Eli Lilly's phase III pipeline. Chorus acted as an alternative path for drug R&D to bring preclinical drug candidates with earlier decision in a "lean-to-PoC" (L2PoC) model to the clinical PoC in a shorter time and at lower costs than the original drug development unit of Eli Lilly. L2PoC refers to the goals to execute clinical studies with limited processing and a focus on delivering the minimum required data package. This "do what is really necessary" principle contrast to the usually existing (traditional) aim of providing data that are "scientifically interesting to have". Chorus defined this as the "quick-win, fast-fail" model of drug R&D.

The main working principles of Chorus are:

- Conduct focused phase I and PoC studies,
- Outsource drug development in a flexible and global model,
- Conduct clinical trials and functional operations like a biotechnology company,
- Manage internal operations and projects with a small internal group only,
- Organize the company in projects with dedicated project budgets, and
- Lean hierarchies and short decision-making processes.

Realizing that the low success rates in phase II of clinical development have the greatest potential to increase the R&D efficiency, the L2PoC-model was better suited to drug candidates that rely on targets with lower target validation and higher inherent technical risks. In addition, it is advantageous if clinical endpoints or biomarkers were already established and the target was related with a focused therapeutic indication. Accordingly, Eli Lilly used the traditional path of drug development for validated targets, in disease areas that necessitated large and long clinical trials to demonstrate PoC, such as Alzheimer's disease, and where deep development, regulatory and commercial expertise was essential.

By 2016, Chorus managed a portfolio of around 15 drug projects and is performing clinical trials in 19 countries, all with a limited staff of 40 full-time employees. To keep the entrepreneurial spirit and to prevent overhead costs, Chorus was built on a flat hierarchy model at which all scientific experts in the Chorus team report to one managing director. In consequence, Chorus was able to run its operation with 25% of its budget as fixed overhead cost and 75% of the financial resources allocated to the external costs of the drug projects.

Overall, this virtualization of R&D proved very successful for Eli Lilly, as the improved success rate in phase II (54% Chorus vs. 29% traditional) enabled the increase of productivity using Chorus model by a factor of 3–10 compared with Eli Lilly's traditional clinical development model.

Central to virtual R&D is the ability to "leverage knowledge." Pharmaceutical companies that succeed in implementing this this strategy have a trim and extroverted R&D organization with low overhead costs. Companies such as Chorus, Shire, Protodigm/Fulcrum, Debiopharm, Cita Neuropharmaceuticals, and Endo Pharmaceuticals have all been able to leverage externally generated innovation (in some cases, even their R&D assets have been acquired from external sources) in combination with a predominantly extroverted way of innovation management. For instance, Cita Neuropharmaceuticals managed to discover and develop a drug candidate all the way into phase II without hiring more than 10 full-time equivalents.

Another interesting success story is **Debiopharm International**, a family owned company established in Switzerland in 1979.[1] For almost 40 years it has operated a unique type of business model that is now being adopted also by much big pharma firms. The success of Debiopharm's two main products Oxalitplatin for the treatment of metastatic colorectal, pancreatic and gastric cancer and Triptorelin have provided the financial independence to implement its unique R&D model.

The Debiopharm group consists of four divisions:

- Debiopharm International: developing & financing innovative drugs;
- Debiopharm Research & Manufacturing: drug modeling, formulation & manufacturing;
- Debiopharm Diagnostics: supporting innovative diagnostic companies,
- Debiopharm Investment: managing the groups' assets.

The resulting flexibility allows Debiopharm to invest in promising external opportunities (compounds and technologies) in and around the healthcare industry with a focus on patient benefits. The company has around 350 employees split over two main sites in Martigny and Lausanne (Switzerland), and has innovation (70% of its staff are scientists) and life-cycle management at its core. Debiopharm's business model utilizes its internal expertise in drug development to identify and in-license promising targeted therapies in the field of oncology and anti-infectives (between the phases of lead optimization and phase III) and then develops these assets with the aim to outsource the results to a partner with sales and marketing expertise. The

---

[1]This case study was generously provided by Dr. Nigel McCracken, VP President Translational Medicine, Debiopharm Int'l.

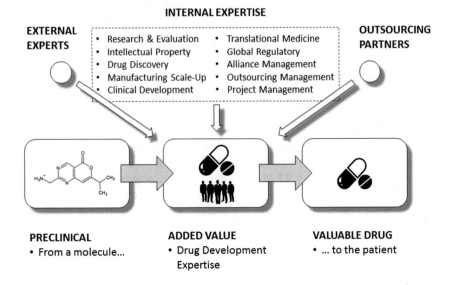

INTERNAL EXPERTISE

EXTERNAL EXPERTS

- Research & Evaluation
- Intellectual Property
- Drug Discovery
- Manufacturing Scale-Up
- Clinical Development

- Translational Medicine
- Global Regulatory
- Alliance Management
- Outsourcing Management
- Project Management

OUTSOURCING PARTNERS

PRECLINICAL
- From a molecule...

ADDED VALUE
- Drug Development Expertise

VALUABLE DRUG
- ... to the patient

Fig. 6.2 The Debiopharm business model

company operates a virtual model engaging external experts and service providers for project-related activities, with the flexibility to involve vendors with latest technologies and solutions to the problem at hand.

The alignment and collaborative nature of the internal expertise within the organization allows Debiopharm to move assets efficiently and effectively through the development process and add value diligently (see Fig. 6.2). Experienced regulatory and market access groups work closely with R&D project teams to conduct and run development programs to support both regulatory requirements (FDA, EMA, etc.) as well and payers' needs. Debiopharm is well placed to be an attractive partner for smaller companies looking for support to have for their compounds guided through development, or also larger pharmaceutical companies looking for co-development partners for one of their portfolio assets, as Debiopharm has a proven track record of drug development without the 'big pharma franchise agenda' which sometimes drives decision making.

Knowledge and collaboration are central to the Debiopharm business model, whether closely working with key opinion leaders (KOLs) and academia, or internalizing knowledge to help better design studies, understand study data or identify promising opportunities. Its expertise in translational medicine helps Debiopharm mitigate development-inherent risks through understanding the mode-of-action, establishing a clinical hypothesis, optimizing dose and study design, and added-value to projects through differentiation, stratification and repositioning. True to its principle of keeping things lean and simple, it has small operative teams focus on connecting the science with the clinical utility, couple with a strong project management group that looks after timelines, key studies and risk assessments. This creates an environment that is flexible and permits quick decision-making.

6.3   Crowdsourcing: Access the Wisdom of Masses

One of the major challenges for the pharmaceutical industry is the access to innovative and validated drug targets as it significantly contributes to the overall low success rates in drug development. Generally, research-based pharmaceutical companies use their internal experts, competitors' information and data from the academia to fill their pipeline of new target ideas. Usually, research scientists provide target proposals as part of their annual objectives. But an emerging alternative path to access new drug targets is to source them from the outside the firm by an open innovation method called 'crowdsourcing'.

Crowdsourcing is an amalgam of the two terms 'crowd' and 'outsourcing' and denotes the strategy and process of an open request of a pre-defined problem to be solved (or knowledge to be generated) to the general public outside the company. Although not specifically conceived for the internet, crowdsourcing is primarily performed with the aid of the worldwide web platform, whereby the crowd is linked to the R&D process by information (posted problem), and the respective solutions are provided digitally (informational crowdsourcing).

Across all industries, **Eli Lilly** is a pioneer in the crowdsourcing field. It founded several crowdsourcing initiatives such as Innocentive in 2001 and YourEncore; both now operating independently (see Table 6.1). **YourEncore** is a global expert network of around 11,000 experts with an average 25 years of industry experience each in technology sectors such as life science, consumer and food industries. The network supports companies to access experts to help to solve the companies' problems in preclinical and clinical development, clinical operations, manufacturing, regulatory affairs, organizational effectiveness, safety, pharmacovigilance, and quality management.

Innocentive is global network of more than 365,000 registered problem solvers coming from nearly 200 countries and problem-posting companies. AstraZeneca, Booz Allen Hamilton, Cleveland Clinic, Eli Lilly & Company, NASA, Thomson Reuters, the Department of Defense, and several government agencies in the U.S. and Europe have partnered with Innocentive to accelerate the generation of innovative ideas and to solve problems quickly. More than 2000 external challenges and thousands of internal challenges have been posted in more than 620,000 project rooms. More than 59,000 solutions were posted since 2001, and more than 2400 rewards have been given so far, with ranging from US$5000 to US$1,000,000 (e.g., the US$1,000,000 for the Amyotrophic Lateral Sclerosis (ALS) biomarker identification by Dr. Seward Rutkove that represented a significant milestone in ALS research).

Other crowdsourcing activities of Eli Lilly are organized on the platform **Open Innovation Drug Discovery**, which includes access to chemoinformatics, biological screening, synthetic chemistry, and high-throughput screening (HTS),

Table 6.1   Overview of crowdsourcing initiatives in the pharmaceutical sector

Name	Originator	URL
InnoCentive	Eli Lilly	innocentive.com
Grants4Targets	Bayer	grants4targets.com
Open innovation drug discovery	Eli Lilly	openinnovation.lilly.com
Your encore	Eli Lilly	yourencore.com
European lead factory	EFPIA	europeanleadfactory.eu
Call for targets	MRC	callfortargets.org

as well as participation in a network for addressing neglected and tropical diseases. This platform helps Eli Lilly access new ideas, drug targets, novel compounds and solve other specific challenges.

The crowd can bring in new ideas, such as target proposals, that are sourced to the R&D pipeline if evaluated positively. The open innovation platforms Grants4Targets (G4T) of Bayer HealthCare and the **European Lead Factory** (ELF), a project of the European Innovative Medicines Initiative (IMI), are prominent examples. ELF is a Public Private Partnership (PPP) that aims at providing new drug targets for all kind of disease areas. Any scientist from a European academic institution or small to medium-sized entity (SME) can bring in a target proposal for HTS or ideas for a compound library. ELF brings together 30 internal partners from private and public organizations, amongst others AstraZeneca, Bayer, and Sanofi. The goal of all participants is to identify high quality small molecule drug candidates. Starting in 2013, ELF delivered 80,000 new chemical compounds in the first 2 years with the goal to achieve 200,000 by end of 2017. Together with the 300,000 compounds that have been provided through the European Federation of Pharmaceutical Industries and Associations (EFPIA) collection, the Joint European Compound Library will comprise 500,000 chemical compounds for HTS screening. As of 2015, 60 drug target proposals had been evaluated positively and accepted and more than 500 hits had been handed over to academic institutions and small-to-medium-sized pharmaceutical companies for further drug development.

In 2009, **Bayer HealthCare** started the crowdsourcing platform **Grants4Targets** (G4T) to identify and validate drug targets and biomarkers. Bayer HealthCare offers two types of grants of 5000–10,000 € (support grants) and 10,000–125,000 € (focus grants). For instance, 125,000 € are given to successful target structure research candidates (Dorsch et al. 2015).

G4T has received noticeable global recognition, as around 2000 interested experts click the website per month, with the majority of the proposals coming from Europe (60%) and the U.S. (23%). Proposals are invited twice a year, with response times of around 2 months. The IP stays with the proposer, limiting the administrative hurdles in the process. If the proposal results in a drug project, both parties can negotiate a collaboration agreement.

In 11 rounds conducted since 2002, more than 1110 applications were filed and 62% were evaluated positively in the first instance. 37% were recommended for recognition after a scientific review, and of those 57% were finally accepted. Most of the target proposals were related to small molecules (63%) in the fields of oncology (64%), cardiology (26%) and gynecology (8%). Half of all proposals were new to Bayer HealthCare. Finally, 147 proposals (13%) were accepted and rewarded with a total sum of 3.2 million € resulting in 6 lead generation, 1 lead optimization and 2 preclinical development projects.

Another crowdsourcing approach is the **PartnerYourAntibodies** initiative, looking for novel antibodies relating particularly to the fields of oncology, cardiology, hematology, ophthalmology and gynecology: all addressing a specific target or pathway in a selective manner. Successful applications are assessed concerning bioactivity, potentially leading to substantial financial funding from Bayer HealthCare. Encouraged by the success of G4T and PartnerYourAntibodies, in 2014 Bayer started yet another open innovation platform in the search for lead structures, Grants4Leads, with 5000 € on offer for suitable submissions. Another website is also planned for suggesting therapeutic indications.

In conclusion, the goals of crowdsourcing are:

• Increase the innovation potential,
• Exploit academic R&D more efficiently,
• Get access to trans-functional know-how,
• Reduce time-to-market,
• Reduce R&D costs, and
• Increase R&D efficiency and effectiveness.

Crowdsourcing provides the advantages of low fixed costs, the ability to access a global brain pool to solve a problem in a fast and non-bureaucratic manner, low organizational hurdles, less hierarchy, and the option to receive a large number of solutions to a posted problem from external experts from both inside pharma and from other industries. In sum, the expectation is that overall R&D efficiency of the pharmaceutical sector will be improved.

The factors for successful crowdsourcing can be summarized as:

• Crowdsourcing initiative needs to be publicly known,
• Submission process needs to be simple and non-bureaucratic,
• Questions and challenges need to be precisely defined, and
• Proposer companies need to be open for external ideas.

Regardless of the obvious advantages, crowdsourcing is a process that is not used yet by all research-based pharmaceutical company, as companies need to face and solve some fundamental challenges:

• Who has the right to file a patent application?
• Who has the right to a patent?
• Who manages the prosecution of the grant of a patent?
• How to manage the not-invented-here attitude of internal R&D?

- How to motivate in-house scientists that may fear to become redundant if crowdsourcing is successfully applied?
- How to circumvent a misconception in senior management that the R&D already employs the best scientists and collaborates with the best experts outside the company?

6.4   Open Source Drug Discovery

It is important to clarify that crowdsourcing is not open source. The open source philosophy is based on transparency, freedom-to-operate, collaborative improvements, and access for everybody to the scientific results and products. There are no legal rights to financial reward for open source contributors; they seek recognition and the satisfaction of providing a better solution to a challenge. Crowdsourcing, however, suits the context of the IP-driven pharmaceutical sector: Crowdsourcing offers a better solution in a profitable-oriented business, as it provides a distinct format of operation and compensation for all contributors. The basic tenets of open source drug discovery are only possible if IP is made publicly available, non-monetary incentives are sufficient for volunteer contributors, and drug development costs are carried by governments, the public and charities.

In pharmaceutical R&D, open source has therefore played a role only in the field of the neglected tropical diseases (NTDs) so far. For example, GSK, Alnylam Pharmaceuticals and MIT have partnered to build the Pool for Open Innovation against NTDs providing open access to 2300 tropical disease patents. In another initiative, GSK, Bayer, Novartis and donors such as the Bill & Melinda Gates Foundation, have founded the Global TB Alliance aimed at discovering and developing better drugs against tuberculosis. Other prominent examples of open source initiatives in the pharmaceutical field are the Open Source Drug Discovery initiative (to provide affordable healthcare for neglected diseases) and the African Network for Drugs and Diagnostics Innovation (ANDI), launched in 2008 to address specific health needs in Africa.

6.5   Private-Public Partnerships

Private Public Partnerships (PPPs) have emerged as another cost-efficient way to pursue drug innovation, in particular in context of NTDs. PPPs are business ventures funded and operated by a partnership of academic or publicly funded institutions and pharmaceutical companies. The principal goals of PPPs are to:

- Look to unmet medical needs,
- Transfer academic research to the market, and
- Improve the industry's competitiveness.

Several PPPs have been established over the past years. Among the most prominent ones are

- The Biomarker Consortium,
- The Critical Path Institute Consortia,
- The Innovative Medicine Initiative,
- The Serious Adverse Events Consortium,
- The National Center for Advanced Translational Sciences,
- Bristol-Myers Squibb's International Immuno-Oncology Network,
- The Global Alliance for Vaccines and Immunizations,
- The Global Fund to Fight AIDS, Tuberculosis and Malaria,
- The Stop TB Partnership,
- The Roll Back Malaria, and
- The FDA-initiated Critical Path Initiative.

Charities also contributed to the discovery efforts, in particular with a focus on neglected diseases, such as the Bill & Melinda Gates Foundation or the Drugs for Neglected Disease Initiative.

For illustration, we briefly describe the **Innovative Medicines Initiative** (IMI). Established by the EU and the EFPIA, the IMI is a PPP that aims to foster pharmaceutical innovation for the benefit of European citizens and to strengthen the competitive position of the European pharmaceutical sector.

With a funding of up to 5 billion €, the IMI is backed to operate until 2024 to create multi-stakeholder cross-functional consortia. One such IMI-initiated consortium is SAFE-T, tasked to identify biomarkers for the early detection of drug-related safety issues to the kidney, liver and the vascular systems. Another consortium is the eTOX project that aims to develop a toxicology database to support *in silico* toxicology analyses. The U-BIOPRED and PRECISESADS consortia are both active in the field of personalized medicine, especially respiratory and autoimmune diseases. Other IMI-supported consortia are as the EPAD consortium, PROactive, PROTECT, and GetReal.

6.6   Life-Science Innovation Centers

Some pharmaceutical companies have set up innovation centers as a driving force behind creativity. New organizational concepts bring scientist from pharmaceutical companies and experts from the academia together to solve problems, provide new solutions and deliver innovative products.

GlaxoSmithKline (GSK) has been a pioneer in this field, starting its Center of Excellence for External Drug Discovery (CEEDD) in 2005. The aim of CEEDD was to better access external assets by a dedicated team of only 20 GSK scientists. As an externally-focused R&D center, CEEDD worked across all therapeutic areas and facilitated drug discovery alliances up to clinical Proof-of-Concept. During its time of operation, the CEEDD team managed 16 partnerships with assets that ranged from discovery to phase II of clinical development. For example, in 2011 GSK entered a US$1.5 billion deal with ChemoCentryx to access their chemokine-based therapeutics and jointly conduct four phase III trials for their drug candidates. But after a chemokine receptor 9-inhibitor did not achieve its primary endpoint in the SHIELD study for the treatment of Crohn's disease, GSK decided to discontinue this partnership. In 2012, GSK closed CEEDD as external collaborations were increasingly operated by non-CEEDD scientists and CEEDD was no longer the primary source of external R&D assets for GSK. Regardless of such setbacks, CEEDD helped to fill GSK's early stage pipeline, and nowadays nearly half of its projects are externally sourced.

In 2010 **Pfizer** initiated the first of eight planned Global Centers for Therapeutic Innovation (CTIs) when signing a US$85 million partnership with the University of California at San Francisco (UCSF). The CTI is part of Pfizer's BioTherapeutics Research Group and was Pfizer's start to implement an open innovation model in its R&D organization with the goal to support translational research between Pfizer and academic medical centers.

Pfizer provides its development expertise and financial and human resources, while the academic partners bring in their research expertise in disease biology, targets and patient populations. The CTI includes more than 80 scientists of Pfizer and combines these resources with principal investigators and post-docs at academic medical centers, such as the Medical Center at Columbia University, the Tufts Medical Center or Mount Sinai Hospital, to progress scientific and medical advances through joint research. Partners in the CTI network include 25 academic institutions, four patient foundations and the National Institute of Health (NIH). So far, more than 500 project proposals have been reviewed.

The goal of the **NIH** collaboration is to identify biologic compounds with activity in a pathway or target of interest to NIH researchers and to Pfizer. NIH scientists identify disease-related pathways or mechanisms as potential drug targets and provide research ideas. The collaboration with Pfizer enables them to move these novel disease targets into therapeutic development using industry-standard translational tools and expertise. For Pfizer, the partnership provides access to NIH's biology expertise. Together, the partners work jointly to move potential drug candidates into the clinical development process. Thus, the collaboration agreement of NIH and CTI include access to NIH's ideas, Pfizer's drug development expertise, publishing rights and resources. A joint steering committee governs the partnership and is responsible for project prioritization and decision-making.

More recently, Bayer HealthCare, the German Merck and Johnson & Johnson (J&J) also started innovation centers. **Bayer HealthCare** opened its innovation center in San Francisco to actively identify and pursue partnerships with academic and biotech researchers in the U.S. **Merck** set up a so-called "fit for 2018 transformation and growth program," with its new innovation center located at the German headquarter a key cornerstone of this change process. The goal is to open Merck's R&D culture towards external innovation and to provide a platform for innovators to translate their ideas into robust and scalable business models. **Johnson&Johnson** started a Life Science Innovation center in San Diego, where its Concept Lab aims to provide open collaboration space for start-up companies so that they can do research before committing additional capital, and the Open Collaboration Space provides desk space to start-up companies and individuals. With both programs J&J provides options for entrepreneurs to get to a preclinical proof-of-concept by using its infrastructure, equipment and resources.

6.7   Rising Importance of Venture Funds

Pharmaceutical companies realize that they are becoming the centers of innovation ecosystems, and that it would be unwise to leave the emergence of attractive outsourcing partners to chance. Thus, they sometimes make financial investments in external service providers with whom they might also cooperate. A good example is the Novartis Venture Fund (NVF), which provides capital for spin-offs in order to reduce non-needed capacities and release entrepreneurial responsibility and capability.

The **Novartis Venture Fund** (NVF) was founded in 1996 with the mission to invest in innovative life science companies for patient benefit, creating attractive returns for entrepreneurs and investors. Today, the Novartis Venture Fund is the largest corporate biotech venture funds with more than US$1 billion under management and more than 40 portfolio companies (with total investments of up to US$30–50 million per company). Including the commitments of other investors, more than US$2.5 billion were invested in NVF companies by end of 2014.

NVF operates from Basel (CH) and Cambridge (Massachusetts, U.S.) but support companies worldwide, especially in the biotechnology/biopharma and medical devices/diagnostics areas. The fund has helped to create more than 1000 jobs around 18 phase I/II clinical programs. For example, the fund was a founding investor in Forma Therapeutics in 2008, a biotechnology company in the field of cancer and other genetically-driven diseases. In March 2014, Forma Therapeutics and Celgene agreed on a US$600 million strategic alliance, with the option for Celgene to acquire Forma later in the process. Forma investors received an upfront payment of US$225 million.

Following the economic crisis of 2008/2009, it was difficult for biotechnology start-up companies to access venture capital (Fig. 6.3). High inherent risks, the lack of ROI and the high management fees of fund managers made institutional investors reluctant. But new investors, such as philanthropic venture funds or corporate venture funds, stepped in, with corporate venture funds alone participating in 35% of all early stage deals in 2013, up from a mere 8% in 2005 (PhRMA 2015). Many pharmaceutical companies complemented their external innovation portfolio (one that was originally build on collaborations and licensing) by corporate venture capital (CVC) to develop an alternative to the less efficient in-house R&D organizations.

In general, pharmaceutical companies use CVC to access new technologies and early-stage breakthrough innovation from start-up companies to mitigate the high

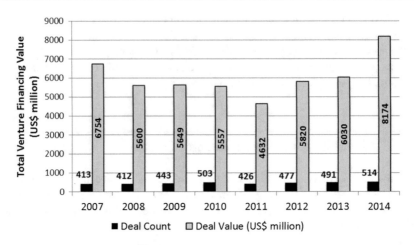

Fig. 6.3 Venture financing deals 2007 to 2014. Source: Evaluate (2015)

risks of very early internal research, while licensing and M&As are used to fill near to mid-term pipeline gaps.

For instance, the **Boehringer Ingelheim (BI) Venture Fund** was founded in 2010 with an evergreen funding of 100 million €. Its goal is to invest in biotechnology start-up companies with innovative concepts and technologies that have the potential to provide breakthrough innovation. Single investments are done of up to 10 million € per company. Key drivers for an investment are managerial and business aspects, as well as the science and hereby in particular the medical need, the innovation potential and the scientific excellence. While BI aims to access drug candidates by licensing and uses R&D collaborations to access technologies with a short-term impact, the BI Venture Fund concentrates on emerging concepts with a long-term impact for the company. The focus of the investments is fairly broad and comprises novel technologies addressing non-druggable targets, new generation vaccines, new generation new biological entities (NBEs), novel projects that enable an entry into new therapeutic areas/indications, technology platforms to identify novel molecular targets/biomarkers, and regenerative medicine/stem cells.

Today, corporate venture funds have been established by most of the international research-based pharmaceutical companies. These funds comprise portfolios of up to 40 companies under management and evergreen funding of more than 100 million € (Table 6.2). Most of the investments are done in the areas of oncology, infectious diseases, central nervous system, cardiovascular, endocrine and metabolic as well as in blood and the immune system.

To summarize, corporate venture funds in pharmaceuticals support start-up biotech companies and other potential outsourcing service companies, to access new breakthrough technologies in a risk-mitigating manner, and to expand R&D activities without investing too much in one single compound as it is done with in-licensing a drug candidate. However, this requires a more complex management

Table 6.2 List of corporate venture funds in the pharmaceutical industry

	Company	Name of fund	Founded in	Funds	Current portfolio (# of companies)
1	Novartis	Novartis Ventures	1996	US$1 bn	40
2	Novo Nordisk	Novo Ventures	2000	US$800 mn	80 since start
3	GSK	SR One	1985	US$1 bn	120 since start
4	Roche	Roche Venture Fund	1990s	CHF 500 mn	30
5	Johnson & J.	J&J Development	1973	US$475 mn	85
6	Pfizer	Pfizer Venture Investments	2004	US$50 mn/y	30
7	Eli Lilly	Lilly Ventures	2001	US$200 mn	19
8	Takeda	Takeda Ventures	2001	US$100 mn	12
9	AstraZeneca	Med Immune Ventures	2002	US$400 mn	12
10	Boehringer-I.	BI Venture Fund	2010	100 mn €	15

of a portfolio of holdings, tax implications, and a more nuanced risk profile of the parent company.

6.8   Managing Intellectual Property Rights

Besides making inventions available to the public, intellectual property rights, patents and trademarks fulfill another important role: It assigns ownership to an individual and thus establishes the legal basis for commercial benefits to the inventor; through enforcement of intellectual property rights, the law protects innovators against imitation and replication of their innovations and knowledge. This protection is crucial in the pharmaceutical industry as otherwise nobody would invest in expensive and long-term drug development. Why should a firm invest US$800 million into a new drug if generics manufacturers imitate it a few months later? Granted, the profits made from successful drugs are impressive; they have led to a demonization of pharma giants akin to an evil empire and garnered substantial popular reaction. However, only few drug candidates ever become successful at all, and it is in the nature of the drug development process and the inalienable principle of drug safety that product development takes this much time and money to complete.

In the pharmaceutical industry, drugs and medicines can easily be copied or imitated because it is not difficult to analyze a pharmaceutical product and determine its respective substances. Due to the significant R&D spending in the pharmaceutical industry and the high risks associated with new drug development, patent protection and the subsequent management of intellectual property is particularly important in this industry. As mentioned earlier, studies have shown that patents are the most effective means of appropriation and found that 65% of pharmaceutical inventions would not have been introduced without patent protection, compared to a cross-industry average of 8% (Reuters 2002). Unless intellectual property is protected with the utmost care, pharmaceutical innovation would not take place as we know it, and overall quality of life (and death) would be significantly reduced. The negative impact for the society would be dramatic.

The 40 leading pharmaceutical firms worldwide have been granted on average 5.8 patents per thousand employees. In 2001, the U.S. accounted for approximately 45% of all pharmaceutical patents that were issued. Japan and Germany both contributed around 10% of patents, followed by the UK with 7% and France with 5%. Looking at company trends, U.S. companies continue to dominate patent approvals in the U.S. The proportion of U.S. pharmaceutical patents issued to U.S. and Japanese companies has increased over the last 20 years, while the proportion issued to European companies has declined: Between 1980 and 1984, U.S. companies were issued around 50% of patents, Japanese companies 13% and EU companies 29%. However, between 1990 and 1994, the proportion of patents issued to U.S. and Japanese companies increased to 55% and 15% respectively, while the proportion issued to EU companies fell to 24% (Reuters 2002).

Leading therapy areas for patent approval worldwide were infectious disease (15%), oncology (14%), cardiovascular disease (10%), neurology (10%), and immune disorders (8%). The distribution of patents across therapy areas largely reflects the balance of the pipeline, and is closely matched to relative unmet need and market opportunity (Reuters 2002).

International patent legislation typically encompasses four statutory classes of patentable inventions that are relevant to the pharmaceutical industry (e.g., Reuters 2003b, USPTO websites): process, machine, manufacture, or composition of matter, with special considerations for living subject matter, especially nonnaturally occurring, nonhuman multicellular living organisms, including animals. Claims directed to or encompassing a human organism are ineligible for patenting.

Process claims refer to the method used to produce a pharmaceutical product rather than to the chemical itself. As it may be possible to develop the same chemical through several different methods, it is often difficult to protect pharmaceutical products solely with process patents or to prove infringement of process patents. Product patents refer to tangible products. Generally, these are commercially viable entities that are ready to be launched or already on the market.

In the pharmaceutical industry, patents are usually applied to medical devices and drug delivery mechanisms, since few manufacturers would want to wait until they perform clinical trials on a compound to apply for patent protection. Depending on the territory of filing, new substances or new processes receive patent protection for a period of 20 years (e.g., by the FDA in the U.S.). However, considering that the average time for drug development in the pharmaceutical industry can reach up to 13 years, the major problem regarding patent protection becomes obvious: the timing of the patent. If a company files for a patent too early, the period when it can market and sell the drug exclusively will automatically be reduced. This is particularly important in light of the fact that pharmaceutical companies only have a relatively short time to market their products and generate a return on their high initial investments. On the other hand, if a company files too late for a patent, it risks losing the invention to competitors. In practice, this means that the effective period of patent protection is rarely more than 8 years in the pharmaceutical industry. Sophisticated methods and techniques to deal with intellectual property are therefore a necessity.

The following case example of **Bayer** illustrates how a pharmaceutical company can utilize and maximize the value generated by its intellectual property by taking a proactive approach to commercialize its intellectual property generated. In this context, Bayer looks at intellectual property as a product of its own. Every intellectual property thus needs its own marketing plan. The intellectual property products are typically spin-offs, 'white-space' developments or technologies (i.e., devices or methods) that are no longer being used by Bayer's business units. When selling the products, Bayer strictly follows the rule, 'don't try to sell any leftovers'.

Bayer developed a four-stage process to decide if certain know-how or a certain technology can be utilized externally (Fig. 6.4). First, Bayer asks if the respective know-how/technology is a surplus product. If yes, the second stage contemplates if the know-how is strategically valuable for any core activity of Bayer's business units. If it is not, the third stage analyzes if the respective technology could be easily brought to a potentially attractive market. If this stage is answered with a yes, the final stage observes if the know-how is not strategically valuable for any other business unit at Bayer. If the intellectual property passes all four stages, it can be marketed outside of Bayer, otherwise it is retained in-house.

Regarding the valuation of the intellectual property, Bayer differentiates between business licenses, product licenses, and technology licenses. The value of business and product licenses, which deal with entire businesses and/or products, can easily be determined by using the scenario technique. The value of technology licenses, however, is much more complicated to estimate and done by looking at the technology maturity and the commercial risk. The combination of both allows for a fairly good estimate. Finally, the marketing plan includes the intellectual property utilization strategy, which could include cross-licensing agreements, royalty payments, cash payments, or equity offerings.

In general, patent protection becomes particularly important in the area of biotechnology. Several ethical questions arise that are not yet covered by existing patent laws and/or acts, including the ownership of genes and whether genes can be patented at all. For example, the Swiss Ethics Committee on Non-Human Gene Technology recently came to the conclusion that intellectual achievements in the area of biotechnology are allowed to be protected. This is justified by the overall purpose of the patent act to support research in the best interest of the public. While it is possible today to receive patents for biotechnological inventions, such as a gene, a genetically changed plant, a biotechnological process or a microorganism, some inventions are excluded from patenting: processes regarding cloning of human beings, processes regarding changes of the genetic identity of human beings, as well as the usage of human embryos for industrial or commercial purposes.

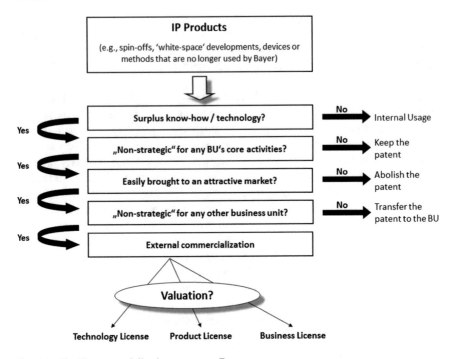

Fig. 6.4 The IP commercialization process at Bayer

6.9   Conclusions

Traditionally, the R&D models of research-based pharmaceutical companies were grounded on the closed innovation paradigm supplemented by individual elements of open innovation, such as research collaborations and outsourcing on-demand. The opening of the R&D process in recent years has increased the output/input-ratio of pharmaceutical R&D, as show the increasing number of drug projects acquired from external sources, the rising importance of collaborations with academic institutions, numerous crowdsourcing platforms, open source drug discovery, virtual organizations, and venture funds.

Some research-based pharmaceutical companies have even aligned entire R&D organizations to access external innovation more efficiently, shifting from a primarily inside-driven R&D to an open R&D model. Eli Lilly established the Fully Integrated Pharma Network (FIPNet), GSK launched its CEEDD and Pfizer developed the CTIs to become more open innovators. Specialty pharmaceutical companies such as Fulcrum established virtual R&D concepts, and Debiopharm used this trim and flexible R&D model to focus on in-licensing and drug development. Such radical open innovation concepts can enhance R&D performance.

The limited implementation and utilization of virtual R&D may be due to some challenges pharmaceutical companies are facing when changing their R&D model from a more closed type of innovator to the knowledge leverager type. The cost related with such a change would be enormous, as are the required changes in strategy, structure, organization, leadership and human resources.

The Internationalization Challenge: Where to Access Innovation

7

> *"There will come a time when breakthrough innovation occurs in China on a regular basis. . . . I don't think there's anything innate in the Chinese that would prevent them from innovating."*
>
> Joe Jimenez,
> CEO, Novartis

7.1 Trends in R&D Internationalization

As a science-driven endeavor, the pharmaceutical industry is inherently global. One the one hand, the fundamental science—biochemistry, molecular biology, genetics, etc.—follows the same natural laws regardless of location on the planet. Especially during the early stages of pharmaceutical innovation, when R&D is mostly a numbers game, the more input there is into the research machinery, the greater the chances of success. On the other hand, the costs of R&D have become so huge that they can be recouped only if they are marketed to as large a population as possible. Drugs sold to twice the number of people cost each patient only half, assuming the same fixed ROI. With governments in most countries having a strong say in the pricing of drugs sold locally, this is an important driver for globalization of R&D as well.

The impetus for globalization is even more acute for pharmaceutical companies from small countries, such as the Netherlands, Sweden, or Switzerland, as they cannot rely on the benefit of domestic homecourt advantage. It is therefore no surprise that the pioneers of R&D internationalization are high-tech companies operating in small markets and with little R&D resources in their home country, as it is the case for Novartis, Actelion, Serono and Roche (Switzerland), Janssen or Solvay (Belgium) or Astra, Pharmacia or Ferring (Sweden). These companies increasingly conducted R&D in foreign research laboratories. Overall, Swiss, Dutch and Belgian companies carried out more than 50% of their R&D outside their home country by the end of the 1980s. Companies such as Pfizer and Eli Lilly

© Springer International Publishing AG, part of Springer Nature 2018
O. Gassmann et al., *Leading Pharmaceutical Innovation*,
https://doi.org/10.1007/978-3-319-66833-8_7

in the U.S., Takeda in Japan, and Bayer in Germany had large home markets and a substantial domestic R&D base, and hence had less pressure to internationalize their R&D activities.

Only in recent years, starting in the mid-1990s, increased competition from within and outside their industries forced companies from large countries to source technological knowledge on a global scale. Global mergers ensued, expanding the research of R&D organizations quickly into new territories. The rise of China as an emerging market for health care and pharmaceuticals has attracted much pharma R&D investment in the mid-2000s. As a result, the top pharmaceutical companies today have R&D locations in all major markets, not just for the coordination of local clinical development projects but also for discovery research.

Drivers and Barriers to R&D Internationalization

The management of cross-border R&D is characterized by a significantly higher degree of complexity than local, single-site R&D management. The extra costs of international coordination must be balanced by synergy effects such as decreased time-to-market, improved effectiveness, and enhanced learning capabilities. Top corporate managers are confronted with the task of finding the optimal R&D organization based on the type of R&D activities, the present geographic dispersion of subsequent value-adding activities such as production and marketing, and the coordination between a multitude of contributors to the R&D process.

The case for R&D internationalization is not unchallenged. Besides the ubiquitous cost argument, foreign R&D units are more difficult to manage, and control, and may be less efficient due to missed scale effects. Table 7.1 summarizes some of the most cited arguments against international R&D.

What drives R&D internationalization? Most factors are due to either science & technology-related issues or sales & output efficiency (Table 7.2). Science and technology-related factors are concerned with R&D personnel qualification, know-how sourcing and regional infrastructure. These factors are largely outside the direct influence of R&D but necessary for its fundamental operations, such as the proximity to universities or the R&D environment. Proximity to markets and customers, improvements of image and collaborations are notable sales & output efficiency-related factors. Efficiency-related criteria focus on the costs of running and the critical mass of R&D units, as well as efficient hand-over processes between R&D and other corporate functions. Direct cost advantages (such as the often-publicized labor costs) rarely influence the internationalization of R&D in the long run, but other efficiency-oriented factors such as costs of coordination and transfer, and critical laboratory size do have an even bigger impact on international R&D organization. Direct costs may become more important in the coming years as the other factors improve in low labor cost countries.

Political and socio-cultural factors such as local content rules, technology acceptance and public approval times, all play an important role in locating R&D abroad. Protectionist, legal and cultural constraints imposed by national governments,

Table 7.1 Barriers to R&D internationalization

Factors in support of central R&D	Obstacles to international R&D
• Economies of scale (critical size) • Synergy effects • Higher career potential • Minimal R&D costs and development time • Better control over research results • Communication intensity • Legal protection • Global product standards • Common R&D culture • Harmonization of regulatory environment • Improved information and communication technologies	• Immobility of top-class personnel • Critical mass (for start-ups) • Redundant development • Language and cultural differences • Effective communication difficult • Much of scientific and technical information worldwide available by Internet • Specific know-how easily lost when support not present • Political risks in target country • Establishment and running costs • No wage advantages in triad nations • Coordination and information costs

Source: Boutellier et al. (2008)

Table 7.2 Reasons for locating R&D abroad

Science & technology	Sales & output efficiency
• Availability of scientists and engineers • Tapping into local scientific community • Proximity to universities • Recruiting local talent • Better R&D environment • Higher quality of life • Lower R&D costs • Higher acceptance for pharmaceutical research	• Smooth hand-over with local marketing and sales organization • Easier coordination with local hospitals during the clinical phases • Compliance with local regulatory requirements • 24-h-Laboratories • 'Good citizen' argument • Local content rules • Protectionist barriers • Tax optimization

Source: von Zedtwitz and Gassmann (2002)

however, often require a company to establish local R&D units. R&D-external forces such as a business unit's striving for autonomy and the build-up of local competence alters the original mission of a local R&D unit. This evolution may take place without HQ's knowledge, particularly in strongly decentralized companies.

In the pharmaceutical industry, mergers and acquisitions have significantly contributed to the internationalization of R&D, particularly with recent cross-border mergers. Even though small size is a driver for internationalization, the global marketing reach of large firms facilitates the leverage of global R&D and innovation. For instance, when Pfizer acquired Wyeth and Merck acquired Schering-Plough in 2009, on paper the two merged firms would have accounted for 51% of the total U.S. pharmaceutical R&D spending and 39% of total world-wide spending (Comanor and Scherer 2013). Subsequent cost reduction programs eliminated R&D units in the combined companies, but since both domestic and foreign units were closed down, this had little effect on their new extent of R&D internationalization. Rather, we observed a centralization of R&D in certain internationally leading regions of innovation (centers of excellence). However, mergers are rarely driven by scientific or technological reasons. Access to new markets and economies-of-scale effects are primary drivers for mergers. Nevertheless, the resulting R&D conglomerate must somehow come to terms with a new and more international organization.

The development of local products requires the early involvement of market and customer application know-how, which is more likely to be found in regional business units. Companies with local R&D exhibit an inclination towards over-emphasizing different local market specification in order to support local autonomy and independence from the parent company. Host country restrictions, such as local content requirements, tolls, import quota, and fulfillment of standards, can attract R&D into key market countries (pull regulations): Both the U.S. and China have plenty of such regulations attracting inbound R&D investments from overseas. On the other hand, home country restrictions may induce companies to move R&D abroad (push regulations): European regulations caused biotechnology R&D to be transferred to the U.S., and lack of homegrown talent forced Swiss pharma companies to set up R&D centers abroad.

Locations of Pharmaceutical R&D Around the World

A study of 9452 R&D sites across various industries (including automotive, engineering, electrical, IT, software, food, chemical and pharmaceutical companies) produced the following overall results concerning international R&D locations (see also von Zedtwitz and Gassmann 2016):

- R&D is concentrated in the triad regions of Europe, the United States, Japan, as well as major regional centers in South Korea, Singapore and other emerging economies in Asia-Pacific, such as India and China. China has seen a particularly strong increase of inbound R&D investments in recent years, but domestic companies are also conducting more and more R&D. Compared to China and India, the other frequently mentioned BRIC countries—Brazil, Russia, and some-times also South Africa—are far behind. They lack economic and institutional attractiveness, and their domestic high-tech ecosystem is still underdeveloped, comparatively speaking.
- Over 70% of all research sites are in the five regions of the Northeastern USA (New Jersey, New York, Massachusetts), California, the United Kingdom, Western Continental Europe (in particular Germany), and the Far East (Japan, South Korea). China (especially Shanghai and Beijing) and India (especially Mumbai and Bangalore) are catching up as locations for research, too. The trend of research concentration is even more apparent when only foreign research locations are considered: almost 90% are in the triad regions of the U.S., Europe, and Japan.
- Although the main regional centers for development largely coincide with the regional centers for research, development is more evenly distributed among European countries and the Northeastern United States, and extends into

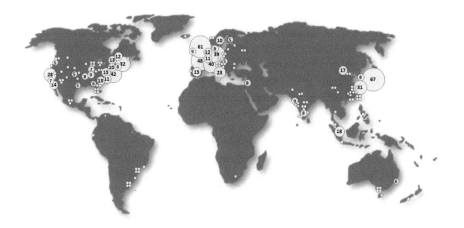

Fig. 7.1 Locations of 821 R&D labs of 45 pharmaceutical firms. Source: GLORAD Database 2016

Southeast Asia, Australia, Africa, and South America. Only slightly more than half of all development sites are located in the eight most development-intensive countries.

Moreover, research and development sites of the same company are not necessarily co-located. This is especially true in the pharmaceutical industry, where development mandates differ so much from research. Since the late 1990s many large companies have made multiple efforts to consolidate their activities in order to realize synergy and coordination potential in international R&D. Transnational R&D projects are managed more easily if the R&D network consists of competence centers (pharmaceutical firms tend to be ahead in reorganizing themselves around competence centers), given that complementary competencies are provided locally. With increasing complementarities of resources, competencies, and knowledge bases, as well as the division of labor and specialization of work, synergy potentials in R&D projects can be exploited.

Figure 7.1 shows a subset of 821 pharmaceutical R&D centers of the 9452 R&D locations. It presents the R&D locations of 45 pharmaceutical companies, including Abbott, AstraZeneca, Boehringer Ingelheim, J&J, Eli Lilly, Eisai, GSK, Pfizer, Merck, Novartis, Novo Nordisk, Pfizer, Roche, Sanofi, Sinopharm, Takeda, or Teva. The distribution of these R&D sites shows a similar pattern to global R&D dispersion across all industries but differs in the following observations:

- There are very few R&D centers outside the globally most attractive countries (the Triad countries, as well as China).
- In the US, R&D is focused more on the North-East, in particular the Tristate area, North Carolina, Massachusetts, and a few centers in the Mid-West. Relatively

speaking, there is less R&D in California, but the Bay Area, San Diego and Los Angeles are catching up also as hosts for pharma and biotech R&D.

- In Europe, R&D is concentrated in the U.K., France, Germany and Switzerland, mirroring the relative strength of these countries as homes of big pharma (Novartis and Roche are from Switzerland, Sanofi and Servier are French, AstraZeneca and GSK are from the U.K., and Bayer and Merck KG are German).
- Despite what some call Japan's lost two decades, there is a strong representation of R&D in Japan, led by the many Japanese pharmaceutical firms, although most of them are smaller compared to their Western competitors (e.g., Takeda or Eisai).
- Emerging new centers of pharma R&D are Shanghai, Beijing and Singapore, but also India (Mumbai and Bangalore), and Israel.

Given this data, the pharmaceutical industry is one of the most internationalized in terms of R&D locations. Pharmaceutical companies also seem to internationalize research as fast as development (albeit for different reasons). Most other industries tend to keep research at home and localize development. Although not obvious from location data alone, pharmaceutical companies also tend to organize R&D as competence-based networks, as opposed to R&D hubs (e.g., automotive and chemicals), polycentric networks (e.g., local market-dependent companies such as Royal Dutch/Shell) or centralized R&D (e.g., dominant design industries). In competence-based R&D networks, each R&D node has a clearly defined competence—and responsibility!—which it brings into the network of other R&D centers. The coordination and management of such R&D networks is more demanding and costly than rather centralized and directive R&D hubs, or the laissez-faire style of polycentric R&D networks. As a consequence, pharmaceutical companies (and many electrical and IT companies, who also favor this type of R&D organization) try to coordinate R&D activities across multiple levels, including the deployment of transnational project teams, platform management, informal as well as formal network techniques.

Examples of Global Pharma R&D: Novartis, Chugai, Ferring

The history of **Novartis** has been documented in detail by Zeller (2001). We will focus here only on some of the more recent developments as they are representative of big pharma. In 2002, about 3000 scientists worked in 10 research centers worldwide in several therapeutic areas. Only 1400 of them were employed in research centers in Switzerland, along with a comparable number in pre-clinical and clinical development: More than half of the R&D workforce was located outside Switzerland. Novartis had 67 drug candidates in clinical development at that time.

In May 2002, Novartis announced it would move its global research headquarters to Cambridge, Massachusetts. Over the following 15 years, the Cambridge-based center has grown by 2000 people in R&D alone and now occupies multiple sites. In 2003, it opened a research center focused on dengue and other tropical diseases (the NITD Institute for Tropical Diseases) in Singapore. It closed its Palo Alto-based

SyStemix R&D and moved it to East Hanover, New Jersey. In 2006, Novartis opened a R&D center in Shanghai, being one of the first of the big pharma to do so.

Over the next few years, Novartis would continue to consolidate R&D in its preferred research hubs. In 2013, Novartis closed its Horsham, U.K., respiratory research center, its center for topical dermatology treatments in Vienna, and a biotherapeutics development unit in La Jolla. It also relocated oncology R&D from Emeryville in California to its Cambridge R&D center.

In 2016, Novartis closed its Schlieren, Switzerland-based ESBATech R&D center, and moved its Institute for Tropical Diseases from Singapore to Emeryville, although Singapore remained a major R&D hub for Novartis in Asia-Pacific. It also shut down its biologics group in Shanghai. However, it opened a brand-new US$1 billion research center in Shanghai, to become its third largest research center behind Cambridge and Basel. As of 2016, Novartis employed 6000 scientists in its global R&D organization, almost half of them in Cambridge alone. Novartis's headquarters remained in Basel, Switzerland.

Chugai is one of Japan's pharmaceutical pioneers. It was acquired by Roche in 2002 but continues to operate independently. Founded in 1925 in Tokyo, it had established R&D centers in Ukima, Kanagawa, and Shizuoka. With new ambitious research targets in the early 1980s, it started to establish overseas offices in the mid-1980s, and to invest or acquire U.S.-based biotech firms. In 1992 it set up its first international research center in Korea, and expanded its London office into a fully-ledged pharma center in 1995. In the same year, it founded a research center in Ibaraki, and 6 years later one in Tsukuba and in Singapore. With the alliance with Roche in 2002, global R&D was reorganized as well. The Tsukuba and the Takada centers were both closed, but the Forerunner Pharma Research company formed, as was a new Clinical Research Center organization, which eventually also opened a base in Singapore in 2011.

An interesting case of internationalization is **Ferring Pharmaceuticals**. Founded in 1950 in Malmö, Sweden, it later moved its headquarters to Copenhagen, Denmark, and then to Saint Prex, Switzerland. It established R&D centers in San Diego in 1996, Tokyo in 2001, Mumbai in 2007, as well as Glasgow and Beijing in 2011. In the U.S., Ferring opened a drug development center in Parsippany in New Jersey in 2008. It also acquired an R&D lab in Israel in 2005. Its main R&D hub is in Copenhagen. In Switzerland, it has an R&D center in Allschwil near Basel.

These examples are typical for the changes in global R&D in the pharmaceutical industry (see Fig. 7.2 for Takeda). We see three major trends:

1. Consolidation in hubs with abundance of talent and technology.
2. Acquisitions of smaller competitors and biotech firms expand the geographic dispersion of R&D.
3. Increased presence in Asia-Pacific, especially China.

The last trend is also observed for global R&D in other industries, which nevertheless requires an explanation, as China is not known to be a place in which

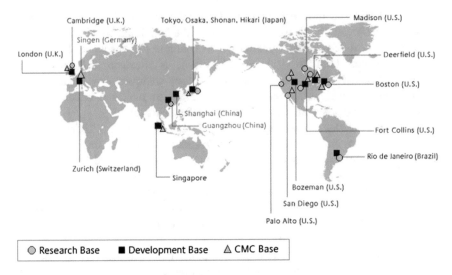

Fig. 7.2 The global R&D network of Takeda. Source: Takeda website 2015. CMC (Chemistry, Manufacturing and Control) research supports compound characterization, process development, pharmaceutical manufacturing, and analytical testing

R&D is conducted easily, nor is it a country in which intellectual property—the main result of R&D work—is well protected.

7.2   New Opportunities for Drug Development in China

The Healthcare Context in China

China is no longer a mystique country closed off to foreigners, but it is still very much misunderstood. Under almost constant reform mode, the China's economy was opened up in the 1980s, picked up speed in the 1990s when it became a major destination for manufacturing investments, excelled at growth rates beyond 13% and 14% in the 2000s, when it also became attractive for R&D, but in the wake of the economic crisis of 2008 it slowed down to what is now called a "new normal" economic growth of 6–7%. It passed Japan as the second largest economy in the world in 2010, and in 2015 its GDP was US$11 trillion, behind the U.S. at about US$18 trillion.

Despite a growing middle class (whose definition is ambiguous in China), most of China is still a developing country, which translates into great tensions about healthcare services needed and wanted, and what role the pharmaceutical industry can play in this context. Improved living conditions further mean that China's population is aging fast, with the share of the population aged 65 or older to double to about 13% by 2025. China's one-child policy, although relaxed to a two-child policy in 2015, led to wealth being distributed among fewer children and family members, which means that more earnings are disposable for life-style and health-related products. Furthermore, as China westernizes, its disease profile will become more and more comparable with the Western world, including afflictions such as hypertension, cancer, diabetes, etc., for which multinational companies already have developed drugs and therapies.

With a population of more than 1.3 billion, China presents a huge market opportunity as it is. A rapidly aging population (in 2015, an estimated 131 million people are aged 65 or over), coupled with new afflictions resulting from air and water pollution, pushes the emerging healthcare system to the brink, and the solutions required will need to concerted effort of government, pharmaceutical industry, and healthcare institutions. Administrative hurdles, low healthcare spending, the lack of intellectual property protection and the poor distribution network infrastructure remain the biggest challenges.

Major pharmaceutical companies pursue aggressive growth strategies and try to benefit from the Chinese market in the longer term. They invest in the sourcing of active ingredients, research and development, and the production and selling of generic and proprietary drugs. In 2002, China was still a low-tiered market of about US$6 billion in total. Pharmaceutical sales grew at a CAGR of 26% between 2007 and 2010, exceeding overall growth in GDP and overall healthcare expenditure (Deloitte 2011). It passed US$100 billion in 2015 and is estimated to grow to US$200 billion in 2020. China has the world's fastest-growing over-the-counter (OTC) drug market and is already the second largest pharmaceutical market worldwide.

Chinese Pharma

Pharmaceutical sales amounted to 17% of total health expenditures in 2015, or US$78 per person. Generics accounted for 64% of total sales, patented drugs only 22% (ITA 2016). The generics market is dominated by low cost domestic producers. Webber (2005) put their number at approximately 6800 Chinese pharmaceutical companies, of which 5000 produced medicines (small-molecule generics and bio-tech products, i.e. not including herbal medicines or traditional Chinese medicines) and the remainder were involved in pharmaceutical-related activities such as packaging and equipment supply. A large share of them are merely producers of raw materials (Wang and von Zedtwitz 2005). By 2015, this market was still highly fragmented, with 5000 domestic companies in operation, of which the top 100 accounting for only one third of the total market (ITA 2016). Government and Chinese FDA (CFDA) policies have been expected to weed out those not in compliance with GMP (good manufacturing process) practices, but have made only limited progress. The 2010 Strategic Emerging Industries initiative plans to create national champions, and policies are being put in place that seem to favor domestic over foreign companies (Prud'homme 2016; ITA 2016).

Chinese manufacturers are very strong in their ability to copy foreign drugs, sometimes selling them under foreign labels. The vast majority of the 3000–4000 pharmaceutical products manufactured in China are copies of foreign products, either legal generics or illegal counterfeits. They serve a worldwide counterfeit market estimated to exceed US$24 billion in 2014 alone (Philipp 2014), with 79% of counterfeit pharmaceuticals seized by U.S. authorities in 2008 coming from China. The situation is complex to control: In addition to the almost 5000 Chinese companies involved in medicine manufacturing, there are 400,000 retail pharmacy shops, and 29,000 firms involved in shipping medical products (Philipp 2014). China lacks effective regulatory control over the manufacture and distribution of active pharmaceutical ingredients (APIs). Chemical manufacturers only need to register with the CFDA if their product is intended for medical use. The CFDA has no authority to regulate them if they declare otherwise, and the CFDA does not monitor or inspect APIs intended for export (ITA 2016). For many Chinese companies, the term 'R&D' refers mainly to the production of additional generic products particularly for China—in terms of strengths, dosage forms and even specific compounds (Webber 2005). As of 2016, what constitutes a "new drug" was still of unclear definition and subject to policy debates (ITA 2016).

In 2015, Tu Youyou won the first Nobel Prize awarded to a Chinese in Physiology or Medicine for her discovery of a novel Malaria therapy, artemisinin. While Traditional Chinese Medicine (TCM) has been a staple in China for a long time, it put the spotlight on the uneasy relationship between western and Chinese approaches to medicine (many Chinese doctors prescribe TCM side-by-side with western ethical drugs). More than 8000 (including different dosage forms) traditional Chinese medicines (TCM) are manufactured and sold. With few market entry barriers, there were over 2500 TCM companies in China. TCM accounts for more than 20% of sales and is particularly strong in rural areas. TCM drugs make up 39% of the Essential Drug List (EDL) and 46% of the National Drug Reimbursement List (NDRL) (Export 2016).

In the Chinese view, TCM has already proved itself to be effective in curing many kinds of diseases, having "gone through thousands of years of clinical trials" (Guo 2016, quoting Wang Guangji, a deputy to the National People's Congress (NPC) and former vice president of China Pharmaceutical University who specialized in western medicine): "But important terms like Yin (substance) and Yang (energy), whose balance are essential for harmonious operation of the body, according to traditional Chinese medicine, would hardly be accepted by foreigners."

Western pharmaceutical firms shy away from TCM R&D, as it is difficult to isolate single, scientifically testable compounds (Waldmeir 2015). According to Ye Yang, deputy director of Shanghai Institute of Materia Medica, "to get a single compound out of a single plant, like in the case of artemisinin, took so many years. In traditional herbal medicine [you might have] 500–600 compounds working together. We don't have the technology to follow all of them to see how they are working individually and together."

The Chinese government promotes TCM, and several government departments are guiding the TCM industry's attempts to modernize. According to Webber (2005), the Chinese authorities have identified two tracks for developing TCM R&D. The first is through purification and standardization to meet global standards and remove impurities such as pesticides and heavy metals. The second track is to utilize TCM as a starting point for producing *novel* medicines. This may be through the identification and purification of the active element(s) (often complex molecules) or the discovery and development of small molecules which mimic the activity of the original TCM. Liu and Xiao (2002) found that about 140 new drugs have originated directly or indirectly from Chinese medicinal plants by means of modern scientific methods.

The Chinese government has also set clear focus on certain areas of biotech research, especially genomics research. In 1998, the Ministry of Science and Technology established the Chinese National Human Genome Centre based in Beijing and Shanghai, and the Beijing Institute of Genomics, as centers of excellence for genome sequencing and analysis, thus enabling China to join the International Human Genome Sequencing Consortium in 1999, in which China played a significant part. China now is on par with major international research leaders in areas such as gene mapping, transgenic technology for animals and plants, gene therapy technology, stem cell research, gene chips and gene research of some major diseases (Webber 2005). The country has several world-class scientific biomedical institutions—the North and South Genome Centers, the Institute of Materia Medica, Tsinghua and Beijing Universities, for example. China also has some 200 research institutes for biotechnology and more than 30 of the 150 key state laboratories in biopharmaceutical-related areas.

In China, the domestic pharmaceutical research and development environment is still dominated by universities and scientific institutes rather than pharmaceutical enterprises. The industrialization of pharmaceutical research and biotechnology still lags the Western world, but is catching up quickly as well (Shi et al. 2014). For instance, **Beijing Genomics Institute** (BGI) was founded in 1999 in Beijing as a non-governmental independent research institute in order to participate in the Human Genome Project as China's representative. It moved to Hangzhou when the local municipal government picked up the expiring HGP funding. In 2002, it sequenced the rice genome, and in 2003 it decoded the SARS virus genome and developed a detection kit. It moved to Shenzhen as the first citizen-managed, non-profit research institution in China in 2007, and in 2008 published the first complete human genome of an Asian individual. In 2010 it established offices in Cambridge (Mass.), and Copenhagen. In 2013 it worked with 17 of the top-20 pharmaceutical companies, and later that year bought Complete Genomics, based in Mountain View, California. In addition to research offices in all the major cities in China, it also has research collaborations or centers at UC Davies in California, Vancouver, Philadelphia, and Hong Kong, as well as a genetic testing facility in Prague, Czechia. BGI has lowered the price for a complete human genome to under US$3000, is the world's largest gene sequencer, and is now regarded as one of the most advanced high-tech startups from China.

WuXi PharmaTech is a biopharmaceutical R&D services firm headquartered in Shanghai. Founded in 2000, WuXi established services in synthetic chemistry in 2001, manufacturing process development in 2003, research manufacturing in 2004, bioanalytical services in 2005, service biology in 2006, and toxicology and formulation in 2007. In addition to opening several Chinese facilities, WuXi acquired the U.S.-based AppTec Laboratory Services in 2008, gaining expertise in medical device and biologics testing and facilities in St. Paul, Philadelphia, and Atlanta. As part of its continued internationalization, it also acquired Abgent, a San Diego-based manufacturer of antibodies for biological research and drug discovery, in 2011. WuXi also has R&D facilities and labs in Cambridge (Mass.), Plainsboro, Munich, Reykjavik, and in Israel and Korea. By end of 2016, WuXi had 14,000 employees in 26 R&D sites that served 2000 corporate clients.

Foreign Pharma in China

Foreign pharmaceutical companies have conducted R&D in Chins since the late 1990s, although in the earlier years the type of research done was very low key and part of larger studies that were rooted elsewhere. For instance, Roche had an R&D office in Shanghai long before it opened a dedicated R&D center in 2004. It was not until the mid-2000s that global pharma companies discovered Chinas as a harbor for strategic R&D operations, perhaps a decade after global electronics, IT and automotive firms started to invest in R&D in China. There are several reasons for this relative late arrival in China:

- Uncertain IP protection: China had reintroduced patent law only as recently as 1985, and neither the general public nor the scientific ecosystem was used to developing, protecting, and honoring intellectual property. The courts, the legal system, and the police had little—if any—experience in dealing with IP cases. Chinese employees had little sensitivity for personal safety or safety as a public good. Tacit knowledge and knowledge that was pre-patent was not easy to secure and appropriate. Of course, these conditions affected also other industries, but in pharma—with its extremely long lead times—such unwanted exposure of technical knowledge is very harmful.
- Fragmented nature and complexity of China's market: 3700 domestic companies accounted for 75% of annual sales, of which 95% operated in the low value generics market (Dierks et al. 2013). Most needs are met by local generics, with only between 6% and 7% being patent-protected innovative drugs.
- Price pressure: 40% of China's healthcare budget was being spent on medicine compared with 10–12% in Western countries, leading to considerable political pressure to reduce prices (KnowledgeWharton 2013).

On the other hand, there were also several strong China-specific reasons to consider R&D in China:

- Quick development of a middle-income market demanding medicines for their evolving needs. These needs stem from rising life expectancy, pollution, and enhanced abilities to detect and diagnose diseases.

- Understanding the uniqueness of the genomics and metabolomics of the Chinese population gives competitive leverage to domestic R&D entrants.
- Access to natural products in China as sources for new chemical entities (NCEs). Newman et al. (2003) calculated that about 60% of nearly 900 NCEs could be traced back to natural products. One famous example is Lipitor, a statin first discovered in natural sources with sales exceeding US$10 billion annually.
- Access and research on TCM: China-based innovations include Artemisinin, invented in China using sweet wormwood and hailed as a miracle malaria drug, Sobuzoxan, an anti-tumor drug, and Huperzine A (HupA), a novel alkaloid isolated from a Chinese medicinal herb, which improves memory deficiencies in Alzheimer patients.

Testing the waters for captive local R&D operations often begins with R&D collaborations. GlaxoSmithKline (GSK), for instance, invested over US$10 million in cooperative R&D with Chinese research institutions since the mid-1990s. Novartis started a collaboration with the Shanghai Institute of Materia Medica (SIMM), with the objective of isolating compounds from Chinese medicinal plants for Novartis to further screen and identify lead compounds. After an initial phase and US$2 million in funding, training and equipment, by 2004 SIMM had isolated more than 1800 compounds from natural herbs covering immunology, oncology, diabetes and the central nervous system.

Other braved the uncertain conditions and started dedicated R&D centers early. In 1997, Novozymes opened an R&D center in the Zhongguancun Science Park in Beijing. Costing 10 million €, the facility focused on the customized development of enzymes and processes for the Chinese market. Roche opened a new R&D center for 40 chemists in Zhangjiang High-Tech Park and established R&D alliances with the state-owned genomics centers in Shanghai and Beijing, to conduct research in the genetic predisposition to diseases such as diabetes and Alzheimer's. Local R&D enabled these foreign pharma companies to enter more credibly into a dialogue with authorities and opinion leaders in the country, which ultimately is good for business in China.

By the mid-2000s, the IP situation in China was better understood, other foreign multinationals had been successful with captive R&D in China, and some of pharma's R&D pioneers seemed to succeed with R&D in China as well. As a result, pharma companies started to invest more strategically in R&D in China, in the expectation to access the natural resource base in China, and to tap into an increasingly large body of medical researchers and pharmaceutical scientists. Also, China is also an attractive base for clinical research, given that there were more than 300,000 hospitals and healthcare facilities. Table 7.3 presents an overview of this early stage of R&D setup in China.

In July 2004, GlaxoSmithKline China set up an OTC (over-the-counter) medicines R&D center at Tianjin Smith Kline & French Laboratories, a joint venture funded by GlaxoSmithKline. Its aim: to excel in creating innovative science-based products to meet consumer needs and support the joint venture's vision of becoming the premier OTC company in China. Pfizer set up an R&D center in Shanghai

Table 7.3 Initial wave of foreign pharmaceutical R&D centers in China

Year	MNC	Location	Investment	Objective
1997	Novozymes	Beijing	10 mn €	Enzymes
1998	Roche	Shanghai	40 scientists	Clinical R&D, factory support, genomics
2001	Servier	Beijing	n/a	Develop potential value for TCM
2002	Novo Nordisk	Beijing	n/a	General biotech research
2003	AstraZeneca	Shanghai	First year: US$4 mn	Clinical research; Collaboration with Health & Medical Institute in China; Localize therapeutic methods
2003	Eli Lilly	Shanghai	>100 scientists	Combining different kinds of organic substance for new drugs
2004	Roche	Shanghai	50 scientists	Phase I: Chemical drugs; analyzing compound structures Phase II: TCM & genetic engineering
2004	GSK	Tianjin	16 scientists	20 new OTC products over 3 years
2004	Johnson & Johnson	Shanghai	n/a	Develop medications suitable for Chinese and Asians
2005	Pfizer	Shanghai	US$25 mn	Trial protocol design and assessment
2006	Novartis	Shanghai	US$100 mn, 400 scientists	Cancer & infectious diseases

Source: SMIE Medicine Information and own research

focusing on developing trial protocol design and assessment of trial results, and Novartis built a US$100 million drug discovery research center in Shanghai in 2006. The research facility, which expanded to 400 scientists by 2008, focused initially on discovering medicines to treat cancers caused by infections, which make up a considerable proportion of the cancer cases diagnosed in China. Some smaller biotech firms also started local R&D in China, for instance BiColl in Shanghai. While pharma R&D focused initially on Shanghai and Beijing, other industries explored locations in second-tier cities and interior China, a trend that pharma R&D is following only slowly. Even globally, pharma R&D tends to be more concentrated in few hubs, and there is no reason to assume this should be different in China (Fig. 7.3).

In the years after the global financial crisis China benefited from its domestic stimulus program, but as of 2013, China's GDP growth has marked slowed down. This is especially true for the already well developed provinces along the Pacific shoreline, while many of the central provinces continue to enjoy high growth rates, albeit from a lower base level (according to the National Bureau of Statistics in China). Cities in those provinces—e.g., Chengdu, Chongqing, and Wuhan—are expected to become more attractive as hosts for pharma R&D.

At the same time, foreign pharma companies continued to consolidate and expand R&D in their original R&D locations. **Roche**, which has occupied the same location in Shanghai since 1997, poured in RMB 1.8 billion (about US$300 million) to expand to 19 buildings covering 70,000 square meters and employing 1800 professionals. Shanghai became its third most important strategic R&D site worldwide, with major projects being run from China globally. Construction is under way

Fig. 7.3 Locations of foreign and domestic pharmaceutical R&D centers in China. Source: GLORAD database (as of 2016)

for the Roche Innovation Campus Shanghai, as is a diagnostic manufacturing facility in Suzhou; these projects cost another US$450 million. **Novartis**, which started with a 5000-square meter facility in 2006, invested US$100 million to build a dedicated R&D facility for 400 scientists in 2007. In 2016, it opened a US$1 billion R&D center in Shanghai, banking on the expectation that the Chinese government's own increased spending on biomedical research and training will deliver talented and highly educated researchers. A drug for liver fibrosis was fully discovered in China and brought into the late clinical stages, with hopes running high that it will be a high-performance global drug soon.

AstraZeneca doubled down on its 2003 start in clinical research in China and in 2007 added a US$100 million investment in its new AstraZeneca Innovation Center in Shanghai. The center's initial research mandate was "In China, For China," but it has assumed a broader mission as a full-fledged discovery center focusing on diseases more prevalent in Asia. In 2015 it made available US$150 as an initial investment in a widened collaboration with WuXi Apptec (so renamed after the merger of Wuxi Pharmatech with Apptec Laboratory Services) to produce innovative biological medicines in China, with several hundred million US$ earmarked for further R&D investments in China across the board.

It is important to note that pharma R&D in China is still very much in its early stages, and it is far from certain that the upfront multibillion-dollar investment will indeed deliver the expected payoffs. Research by Grimes and Miozzo (2015) showed that only 1.78% of the 9543 patents granted by the USPTO between 2004 and 2014 by the top big pharma companies had Chinese citations and only 1.44% had Chinese inventors. Of the latter, Roche accounted for 65% and Novartis for 18%. China's share of biotechnology patents was also at around 1%, which is the same level as its overall USPTO share. While China is often credited with speed and rapid development, many of the problems that kept pharma investment away in the 1990s are perhaps still present, or at least inadequately addressed (see Table 7.4). AstraZeneca's

Table 7.4 12 challenges big pharma faces in China

#	Challenge	Explanation
1	China's size and heterogeneity	There is not one China but 1000 Chinas. A level 3a hospital in a coastal tier 1 city is unrecognizable from a rural polyclinic in Western China
2	Public sector complexity	No 'quick fixes', too many stakeholders involved in public healthcare provision
3	Poor patient-physician relationships	Absence of gatekeepers to control public sector referrals, leading to long waiting times for appointments, high patient dissatisfaction, deterioration in doctor-patient relationships and sub-optimal patient outcomes. Medical practice "a high-risk job"
4	Rationing of high cost drugs; patient self-pay	China's public hospitals treat more patients with less budget. Patients often pay much of the cost of treatment themselves, thus deferring treatment until the condition is very severe
5	Undeveloped private sector	Negative attitudes towards private healthcare, which makes it very difficult for the private sector to attract good doctors, who feel the public hospitals offer them better career progression opportunities
6	Slow drug approvals	CFDA backlog: In 2015, over 18,000 drug applications had yet to be reviewed. Worse for foreign pharma: Phase III trials to be conducted in China regardless of where the product was previously launched
7	Medical tourism and gray market imports	Drugs are expensive in China, and often available much later than abroad. E.g., drugs are the most purchased items by Chinese tourist in Japan
8	Scrutiny on international companies	China's anti-corruption initiatives make pharma companies very tentative when operating in China. Compliance with China specific regulations—including anti-commercial bribing and anti-unfair competition law, is critical
9	Preferential conditions for domestic players	China's domestic pharma are not only less subject to the restrictive compliance regulations but also given preferential access to drug lists—such as the National Essential Drug List (NEDL)
10	Limited access to time-poor physicians	Time for physicians to engage with medical reps or educate themselves on new treatments is extremely limited—especially important in a country that values personal connections ('guanxi')
11	TCM Cmpetition	203 of the 520 compounds on the 2012 edition of the NEDL were TCMs
12	Economic slowdown	For patients, economic slowdown means everything costly becomes even costlier. For foreign pharma, RMB devaluation means China is now a cheaper market to invest in, but also a less attractive one

Source: Marc Yates, eyeforpharma, 16 Nov 2015

former head of R&D in Asia noted that because of China's shortage of experienced toxicologists, pathologists, statisticians and clinicians, it could take several decades before China's pharmaceutical ecosystem was fully developed (McKinsey 2012).

The question is whether economic growth coupled with the shaping of a positive environment will encourage increasing investment into innovative pharmaceutical R&D in China. The global pharma companies are well advised to explore the China option carefully and systematically, given the dangers involved, but certainly they are advised to take this opportunity seriously.

7.3   Reverse Innovation in Healthcare

Reverse innovation is a new paradigm of global innovation describing the reverse flow of innovations from emerging countries to the established industrialized markets of the United States, Europe and Japan (Govindarajan and Ramamurti 2011). It captures not only innovations first introduced in such markets but also those developed there (von Zedtwitz et al. 2015), and it also includes truly indigenous sources of innovation rather than primarily foreign-invested subsidiaries of technology-intensive multinations in those regions (Corsi and von Zedtwitz 2016). While often associated with low-cost solutions to problems in high-priced markets (Zeschky et al. 2014), reverse innovations are not just cost-efficient products gone global but create value-added benefits specific to targeted market segments in both emerging and mature markets.

Reverse innovations have quickly attracted the attention of healthcare and pharmaceutical experts (Syed et al. 2013), primarily because healthcare costs have been escalating in advanced country markets, and emerging market countries solutions promise a more cost-efficient containment approach to many healthcare-related problems in the West (Crisp 2010). As the largest emerging country, China is the "usual suspect" as a source of such reverse innovations, but other countries and even continents (e.g., Africa) are making substantial contribtuions as well (see e.g., Harris et al. 2015). India has already developed a reputation for excellent yet affordable healthcare (e.g., the famous Aravind Eye Hospital), adding to its growing expertise in frugal innovation (Weyrauch and Herstatt 2016).

The list of examples illustrating the potential that reverse innovation can have on healthcare in general and the pharmaceutical industry in particular is growing. Among the first instances of reverse innovation mentioned ever are medical devices: GE's Mac400 Electrocardiogram from India, or GE's portable ultrasound devices developed in China. But medicines are also increasingly coming from emerging market countries: Vicks Honey Cough drops were developed by P&G in their Caracas R&D center, initially for their Latin American market, only to see it do well with large populations in the U.S. and Europe as well. Oncovin is a drug that Eli Lilly developed in the U.S. based on active ingredients found—and initial research done—in Madagascar. Most famously perhaps is Artemisin, a key ingredient in Chinese TCM, that was developed further as an anti-malaria drug, and ultimately led to the 2015 Nobel Prize for Medicine to be awarded to a Chinese researcher, Tu Youyou.

Innovations from emerging markets carry the stigma of being 'low cost' or 'geared towards the needs of under-developed civilzations' (for a summary of roadblocks against reverse innovation, see Hadengue et al. 2017) and often face steep adoption and acceptance difficulties (DePasse and Lee 2013). Western policymakers and governmental institutions have therefore launched several initiatives at making the reverse transfer of emerging market solutions (from so-called LIC low-income countries) more predictable and more systematic:

- Prize competitions: Snowdon et al. (2015) reported the results of an open competition for healthcare solutions from developing or emerging countries, with the winner to be awarded US$50,000 for further financing their project. A committee evaluated all 12 submitted projects on the basis of seven weighted criteria. Apart from the winning innovation being supported through the prize money, the competing entries also benefited from project-specific feedback, and through media exposure the public was made aware of the potential of reverse innovation in health care.
- Crossover identifications: DePasse and Lee (2013) described a four-step model to transfer reverse innovations that includes a common problem identification in a LIC, a dissemination analysis in that country, a crossover proposal, and subsequent dissemination in a high-income country (HIC). The trick is not only to execute that crossover effectively, but also to identify suitable reverse innovation candidates and to support disseminiation in the home country.
- Scoring assessments: Battacharaya et al. (2017) developed a two-step scoring system with eight assessment criteria, mostly for use by policymakers, public institutions or funding agencies. The purpose of this assessment is to identify potential reverse innovations and qualify them for their suitability to be transferred and applicability to an unmet need in the home country.

In conclusion, the potential of reverse innovation in pharmaceuticals is still difficult to estimate, but the healthcare sector has taken a strong interest in considering any emerging solution from emerging market countries due to the mounting cost pressures in industrialized nations. Pharmaceutical multinational firms that already have R&D centers in those emerging markets are well positioned to leverage their reach into different population groups not only as patient pools but also as sources of innovation, and ultimately may play an important role in acting as conduits of reverse innovation in their home countries.

7.4   Conclusions

R&D in multinational companies has developed from centralized and geographically confined towards distributed and open structures. The pharmaceutical industry is one of the most advanced in terms of R&D internationalization, and one of the most specific when it comes to regulation and significance of science and technology. Still, maintaining a well-balanced locally responsive and globally efficient R&D network is one of the great challenges of multinational organizations. Rapidly evolving biotech firms and new pharma entrants (whether from emerging economies or in the U.S., Europe or Japan) repeat those growing pains that big pharma companies such as Novartis, Pfizer or Takeda have experienced since the 1980s.

The key lessons learned for managing global R&D in the pharmaceutical industry can be classified as follows:

- Localization of management resources;
- Flat and flexible organizations;
- Introduction of local culture of innovation and know-how;
- Challenging projects coupled with bottom-up creativity;
- Personal interactions more important in decentralized R&D;
- Synchronization of international drug development by means of transnational project management in order to shorten R&D cycles;
- Worldwide integrated R&D data management;
- Acquisition of external ideas and projects as important as internal R&D;
- International teams require new organizations;
- Manage platforms, not individual R&D projects;
- Better inclusion and leverage of local university and national research programs;
- Effective management of local open innovation partners;
- Foster networking and collaboration.

In conclusion, transnational R&D management is a key ingredient to success, and the high stakes of the drug approval and medical safety have made the pharmaceutical innovation pipeline one of the best understood R&D engines (Gassmann and von Zedtwitz 2003b). However, there is still untapped potential to improve this engine with new technologies, new managerial approaches, and new scientific talent drawn from countries around the world.

Future Directions and Trends

8

> *"The future is already here,*
> *it's just not evenly distributed."*
> William Gibson,
> The Economist

8.1 Three Levels of Future Change

The pharmaceutical industry is at the beginning of a huge transformation; a transformation in which innovation will drive future competitiveness and growth even more. Improving the flow of innovation, increasing R&D efficiency and closing the productivity gap are the central challenges in this next transition. Given major technological changes in discovery technologies, drug targets, design and formats, new market and customer expectations such as individualized medicine and reduced overall healthcare costs, as well as new competitors arising from the digital corner, the pharmaceutical industry is facing nothing short of a revolution.

This final chapter summarizes the key conclusions from the book and identifies some trends and suggests future directions in leading pharmaceutical innovation on three levels:

1. Increasing R&D efficiency and effectiveness,
2. Redefining and rethinking the business model, and
3. Leading people for innovation and change.

All of them are equally important, yet each company will need to prioritize actions and strategies differently and in accordance with their own intended path.

© Springer International Publishing AG, part of Springer Nature 2018
O. Gassmann et al., *Leading Pharmaceutical Innovation*,
https://doi.org/10.1007/978-3-319-66833-8_8

8.2   Level 1: Increasing R&D Efficiency and Effectiveness

Openning Up the Innovation Process

Today's companies already rely on vast networks of various types of external partners to improve efficiency of their R&D organizations. This will be even more critical in the future. Partnerships will be created that go far beyond the well-known technology sourcing strategies represented by early pharma-biotech alliances. Intensive competition, access to markets, scarce own resources, lack of know-how, cost cutting or restructuring, growth aspirations, synergies and efficiencies, and last but not least risk reduction are the paramount reasons why pharmaceutical companies increase R&D collaborations with external partners. This implies huge challenges towards the dynamic capabilities of innovators and managers to develop and manage strategic alliances and networks for a better product pipeline.

More Focus on Commercialization Along the Value Chain

Any strategy aimed at improving the R&D pipeline will need to reiterate the focus on commercialization, even of early research results. While the search for in-licensing candidates is widely accepted, the search for out-licensing candidates coming from own research and intellectual property is not sufficiently enforced. Intellectual property should be actively created and bundled so that it can be marketed subsequently to potential licensees. Due to an expanding number of niche-markets, specialty pharmaceutical companies are tempted to acquire late stage projects without having the opportunity costs and risks at earlier stages. Several potential licencees may have a particular need for the pharmaceutical company's intellectual assets. The pharmaceutical company should not only regard these licencees as potential competitors who possibly 'steal' blockbuster revenues. Instead, these firms should be treated as partners with complementary sales and commercialization capabilities for additional revenues and profits.

New Discovery Technologies to Leverage the Potential of Personalized Medicine

We also see a trend towards the use of high-throughput cost-efficient next-generation sequencing (NGS) technology to identify mutations in the whole genome of individuals (whole genome sequencing, WGS) to predict causes of disease such as cancer. WGS technology is used for high-throughput profiling of patients to match their genetic profile with the profile of investigational drugs. Pharmaceutical companies need to invest in NGS and WGS technologies to use the full potential provided by personalized medicine. Microfluidic or lab-on-a-chip technologies are another technological trend that will drive R&D in the future. These miniaturization technologies will ease the use of bioassays and cell biology in biomedical research, will enable high-volume biological testing in drug discovery, will mimic organs on a chip and will improve overall R&D efficiency. These technologies are expected to deliver exponential performance improvement over time, resulting in radical cost reduction similar to Moore's Law in the computer industry.

New Biologics to Capture the Full Growth Potential

Another important technological trend takes place in biologics, such as chimeric antibodies, bi-specific antibodies or orally available antibodies. For example, antibodies are still administered intravenously or subcutaneously: both approaches are associated with the costs of hospital administration, time investments by both doctors and patients, travel and infrastructure. First evidence suggests that orally available antibodies could be available in the near future and antibodies in a pill would be a breakthrough for patients and payers. The advantages of these kinds of new biologics are reduced costs and an increased patient compliance. The pressure for more convenience and lower costs will continue in health care.

New Drug Technologies to Provide New Blockbuster Drugs

Antibody-drug-conjugates (ADCs) and synthetic or in vitro transcribed (IVT) mRNA-based therapeutics are technologies that have the potential to launch the next wave of blockbusters. For example, ADCs are a new class of drugs that include an antibody and a cytotoxic agent in one entity. They offer the specific delivery of the cytotoxic agent to the target by the antibody part and they are less toxic than conventional cancer drugs. Major research-based pharmaceutical companies (among them e.g., Roche, Pfizer and Sanofi), have projects on ADCs in their R&D pipelines. Other pharmaceutical companies (e.g., Eli Lilly and Novartis) have accessed ADC technologies from biotechnology companies through licensing deals. Market research forecasts that the synergistic effect provided by ADCs may translate into a premium price; for instance, Adcetris (Brentuximab Vedotin by Takeda) has been priced at US$14,000 per regimen for the treatment of Hodgkin lymphoma with an annual total sales potential of more than US$100,000 per treatment.

Integrated Solutions to Treat Complex Diseases

Neurodegenerative diseases are a major burden for healthcare systems already, and this burden is likely going to get worse. Solutions in the treatment of complex CNS diseases cannot solely come from drug discovery but need to be derived from a combination of pharmacological and non-pharmacological interventions. Pharmaceutical companies will need to consider stepping into the medical device business to expand their CNS treatment options with electric, robotic and software-based approaches. As a consequence, pharmaceutical companies will need to change from pure drug makers to integrated healthcare solution providers.

Replacing Disease-Centered Approach with Systematic Early Discovery

We expect pharmaceutical research to be further improved by a wide range of novel drug discovery technologies already in use that go beyond high-throughput screening, combinatorial chemistry or bioinformatics, and increasingly center around molecular drug design. The application of these technologies allows for the automation of much of the discovery effort, promoting a more comprehensive and consistent screening process. This means to move away from a disease-centered to a systematic and mechanistic discovery process in the early phases. Research bottlenecks and oversight based on lack of human attention and processing power can be avoided, or at least compensated for. As a result, both the quality and quantity of resulting lead compounds are expected to increase.

Reducing Serendipity in Discovery by Balancing Data Generation and Data Analysis Tools

The ubiquity of information technologies results in huge amounts of data being generated and collected every day. Following the success of these new information production capabilities, we are still developing the techniques to manage and interpret all the accumulated data. Medicine is ripe for new ways to find knowledge in data sets we already have. Big data will change life sciences and healthcare between patients and providers, and between pharma companies and customers. Advanced database technologies, faster algorithms and improved inductive statistical analyses are necessary to efficiently screen through vast quantities of information possibly eliminating at some point the reliance on serendipity alone for future success.

Focusing on Complexity and Diversity in Compound Libraries

Besides applying more advanced screening methodologies, the quality of the substances to be screened has to improve as well. Therefore, the structural complexity and diversity of the compound libraries must increase. This ultimately allows raising the probability to find substances capable to influence a given target in a specific desired way. The ability to manage complex knowledge—identifying relevance and dealing with information dynamics—will be key for R&D driven pharmaceutical companies in the future.

Overall, R&D efficiency and effectiveness in the pharmaceutical industry increase with the company's abilities to absorb, create, transform and interpret relevant knowledge. The effect in terms of productivity will be similar to the introduction of platform development in the automotive industry. Pharma R&D mainly becomes a knowledge management challenge in which the speed to identify and interpret relevant patterns of knowledge generated inside and outside the company distinguishes the winners from the losers in the innovation race.

8.3   Level 2: Redefining the Business Model

Finding the 'Sweet Spot' in the Pharmaceutical Value Chain

Reduced efficiency and flexibility, difficulties to transfer know-how and an unclear intellectual property situation are seen as major decelerators in pharmaceutical R&D. Leading service provider—already complementing important competencies of pharmaceutical companies—may assume even more important roles in managing pharmaceutical innovation. The make-or-buy decision of a contracting pharmaceutical company, i.e. the balance between in-house and external activities, is mostly competence and know-how driven and not capacity or cost driven. Preferred partnerships and collaborations on a project-by-project basis with preselected vendors are the two most favorable cooperation models in practice. However, the control of the critical components in the value chain remains the key issue.

Becoming a Knowledge "Leverager"

In open innovation, the knowledge leverager is particularly skilled at leveraging external innovation with the subsequent benefit of higher R&D output while reducing R&D costs. In the pharmaceutical industry, Shire, Chorus and Debiopharm are examples for the success of this open innovator type. The pressure to innovate makes it necessary for some pharmaceutical companies—second-tier innovators, mostly—to realign their R&D concepts to a more open and cost-efficient setting. Some companies may even be forced to an even more fundamental change in R&D: the radical knowledge leverager concept with virtual R&D teams and a high proportion of externally acquired pipeline projects.

Developing Downstream Capabilities Within R&D

A stronger market-orientation in R&D is one of the primary tasks of all future R&D activities. This in turn requires a stronger business orientation of research managers and key scientists. This is not easy since 80% of all scientists in the pharma sector will never see the commercial results of their efforts. Many scientists in pharmaceutical R&D still know too little about the markets that their products are expected to serve. Communicating these business benefits to scientists more clearly is necessary. As a consequence, an effective and collaborative interface between R&D and

marketing must be established. Ultimately, a stronger market-orientation is expected
to result in a shift from a product to a patient-driven strategy.

Exploiting the Potential of Emerging Markets

Pharmaceutical companies with reduced R&D efficiencies cannot meet their growth
objectives solely by product innovation, and they turn to growth in generics, in OTC
products and in emerging markets. They need to expand their business models from
purely research-based and focused on traditional markets to more diversified and
balanced with respect to the specific needs of emerging markets.

Strategies for How to Benefit from Scientific Breakthroughs

The more we understand the human genome, the more we will be able to use
validated biological targets as starting points for new drug development. The number
of these targets is expected to grow twentyfold to over 10,000 in the near future. The
integration of genomics, proteomics, molecular design and other technologies will
lead to improved target identification and attrition, enhanced lead optimization, and
improved clinical trial designs that speed up approval. This will signal a shift from
broadly targeted drugs to more focused medicines with much higher therapeutic
value for the target population. But access to genomics technologies is not cheap
(estimated minimum US$100 million annual commitment), top-tier pharmaceutical
companies are likely to be the first to fully integrate these new technologies.

Specialty Medicine as a Scope of Business

Traditionally, medicines for treating diseases with high prevalence and high earning
potential have been driving pharmaceutical R&D pipeline decision making. Torn
between commercial needs and ethical interests, specialty medicine will have a
future once development costs meet lifecycle returns. Especially those specialty
drugs with shorter time-to-market and leaner reimbursement and pricing have
blockbuster potential; the trend toward creating blockbusters in this field will
increase in the future.

Understanding and Exploring the Business of Consumer Electronics

A big opportunity are the new life style trends of wearables. New competitors are
creating a market with great relevance to health care. Apple, Samsung and Google
are developing wearable devices with life style management and accident, injury,
and disease prevention functionalities. Their products do not need much clinical
development or FDA approval. Instead, customer acceptance determines success of
the innovation, and new business models based on individual information are

generated. Several pharma companies have started to exploit these fields but none of them has launched a major market success yet. The dominant innovation logic of software and consumer electronics companies is fast paced, incremental technology and heavily user centric. Pharma companies are challenged to succeed in a market that is in part following consumer electronics logic, but increasingly is part of a new domain in health care.

Overall, pharma companies think in compounds and diseases and less in business models. A business model is the integrated answer to four questions: Who is your target customer? What is your value proposition? How is the value proposition implemented (i.e., what is your supply chain architecture)? And why is the business profitable (what is your revenue model)? Software "eats the world," as Marc Andreessen pointed out in a Wall Street Journal article. Although the ubernization of the economy has not reached the pharma sector yet, a careful rethinking and redefining of the business model becomes more crucial. A business model creates and captures value for the company. The higher the pressure on its profit margins, the more a company needs to think in business models.

8.4 Level 3: Leading People for Innovation and Change

Creating a Mindset of Change

Competitive advantage is only temporary: Sustainable profitable growth is only possible if a company continues to challenge the sources of its success. But the annual McKinsey surveys show that only 30% of strategic initiatives truly succeed while 70% fail. The reasons are known (but often not thoroughly considered): Two thirds of those failures are caused by resistance of employees or not-consistent management behavior. The mindset for change is often missing, and if strategies are to succeed, this mindset has to be developed in a consistent and persistent way.

Developing a Transformation Story

While change is a biological and perhaps universal natural imperative affecting all humans, as individuals we often resist change even if the benefits are apparent. This pattern is replicated at the organizational levels: Most companies show cultural inertia and complacency in the face of necessary transformations. Employees as well as managers ask themselves two kinds of questions: (1) Will this really work? And even more importantly: (2) How will this affect me? What will be my new role and my new tasks? With whom will I work together next? Will I be able to live up to the new expectations, and is my job still secure? A convincing transformation story helps everybody affected to understand the need for change and to increase its acceptance by nearly four times, as research on change has shown. The content of a story can be a strong positive vision ("Win against HIV!") or a threatening mission ("Innovate or die!"). Both kind of transformation stories can work if communicated consistently.

Overcoming Functional Silos

People are at the center of organizations, thus improved people development and training will result in better run companies. Because of the long lead times and the great job specializations involved, pharmaceutical companies especially will have to overcome the distance between early-stage research and the final product by adapting their incentive and job assignment systems. Not every scientist wants to be involved in marketing, but every researcher should have had the exposure to the rest of the business, and the opportunity to pursue a modern, more flexible career. All employees in R&D should ask themselves what their contribution could be across all levels of the R&D process. This approach challenges the linear structured process and places more emphasis on group and team-oriented R&D. Feedback loops (i.e. from the clinical trials back to basic research and the screening stages) must be proactively pursued. Modern drug discovery requires the integration of knowledge from a broad array of disciplines. The formation of multidisciplinary teams, including biologists, physiologists, biochemists, as well as specialists in the traditional disciplines of synthetic chemistry and pharmacology, and more esoteric specialists like molecular kineticists, can help to overcome bottlenecks in creativity and develop unexpected downstream opportunities. Marketing experts should be included as early in the R&D process as possible. The most successful pharmaceutical companies differentiate themselves by actively confronting the tension between a pure functional organization versus an organization along product groups.

Balancing Scale and Creativity

Creativity in pharmaceutical R&D is more likely to occur in small teams rather than large ones. Mega-mergers and the resulting even larger companies endanger R&D's ability to be creative. Hence, as with any acquisition and/or integration of outside knowledge into the inside portfolio (which is expected to be a major driver for improvement in R&D in the future), there is the potential loss of creativity and the occurrence of the not-invented-here syndrome. **Henkel** has established the so-called 3×6 teams: six persons work for 6 months together in an autonomous team; six specific product concepts are expected. The idea of small teams and high time pressure can be transferred to pharma companies. Embedded small functional units within the larger unit will help to retain creativity within the firm.

Fail Often to Succeed Sooner

It is more economical and productive to terminate less prospective projects and concentrate resources on objectively more promising ones. There is a cut-off point after which project management gives discipline absolute priority over incremental improvements. Before this cut-off, the R&D organization should be designed for maximal creativity and effectiveness of its discovery effort. **Google X** deliberately focuses to solve the most difficult questions at the beginning of a project with the goal to learn fast, get external feedback early, and then do the next loop. The more a team fails, the more it learns and has the chance to adapt and succeed. Overall, the combined thrust of all project activities determines a company's competence areas and thus its therapeutic fields.

Remembering That Drugs Solve Global Problems

Pharmaceutical products can be applied anywhere in the world. Hence, R&D increasingly adopts an international outlook. Foremost, it is important to align R&D strategy with corporate strategy. Is global R&D a consequence of business decisions, or is global business a consequence of R&D decisions? In this regard, it is essential to clarify what the global decision criteria are and which criteria would influence the initial mission, ramp-up and evaluation of new R&D projects. Company-wide knowledge management systems and communication technologies play a critical role in the overall internationalization process.

The decline of R&D productivity as measured especially in FDA approvals or NCE released has led to intense criticism of pharmaceutical management. The pharmaceutical industry relies on predictable production of new medicines and therapies. While we have found many threats and risks that endanger the prospects of individual companies, overall the future for this industry is bright and exciting. New drug discovery technologies are in the works and expected to revolutionize the way pharmaceutical companies manage innovation.

In addition, new markets emerge. As life expectancies increase worldwide, the aging populations in established markets and the growing population in emerging markets represent many untapped opportunities. The incorporation of market-oriented aspects, improved human resource and project leadership, better R&D pipeline management including a balanced approach to outsourcing and collaborations, as well as a sound internationalization strategy are important elements in a general strategy towards prosperity for tomorrow's most successful pharmaceutical companies. Management and leadership on all levels play a more important role than ever. In times when innovation and change become more important, focus on people is crucial. The companies that will win the innovation race and therefore the race to sustainable competitiveness in a global scale are the ones that win the war on technology and talent.

Glossary

ADCs ADCs are "antibody-drug conjugates", a new class of highly potent biopharmaceutical drugs designed as a complex of an antibody linked to a biologically active anticancer drug. ADCs are intended to target and kill only the cancer cells and spare healthy cells.

ADME ADME is short for "absorption, distribution, metabolism and excretion" and describes the disposition of a drug within the human body. The four criteria influence drug levels and kinetics of drug exposure to the tissues and hence the pharmacological effect of the underlying medicine.

API An API (Active Pharmaceutical Ingredient) is any substance or mixture of substances intended to be used in the manufacture of a drug. When used in the production of a drug, it becomes an active ingredient of the drug product.

BCR-ABL The abbreviation BCR-ABL is used in the context of the Philadelphia chromosome, a genetic abnormality in chromosome 22 of leukemia cancer cells (particularly chronic myelogenous leukemia (CML) cells). This chromosome is defective and unusually short because of the translocation of genetic material between chromosome 9 and chromosome 22, and it contains a fusion gene called BCR-ABL1.

Biochips While containing DNA, Biochips are used to au-tomate the sequencing of genes.

Bioinformatics The use of IT in pharmaceutical R&D (e.g., elec-tronic databases of genomes and protein sequences, and computer modeling of biomolecules and biologic systems). Bioinformatics is expected to expedite lead discovery by providing structural data and analysis for drug targets.

Bioprocessing The creation of a product utilizing a living organism.

Blockbuster A pharmaceutical product earning annual revenues in excess of US$ 1 billion.

Business Model Conceptualized logic of how a company creates and captures value by addressing four key questions: Who is the customer? What is the value proposition? How is the supply chain architecture organized? Why is it profitable?

CAGR Short for "compound annual growth rate", an annualized average rate of change often used in context of revenue growth.

CDER CDER is the FDA's Center for Drug Evaluation and Research. It is responsible for public health in the U.S. by ensuring that safe and effective drugs are available to improve health of people in the U.S.

Chemoinformatics The combination of chemical synthesis, biological screening, and data mining approaches used to guide drug discovery and development.

Cloning Using specialized DNA technology to produce multiple, exact copies of a single gene or other segment of DNA. The resulting, cloned (copied) collections of DNA molecules are also referred to as clone libraries. A second type of cloning exploits the natural process of cell division to make many copies of an entire cell. The genetic makeup of these cloned cells, called a cell line, is identical to the original cell. A third type of cloning produces complete, genetically identical organisms (e.g., animals).

CMR CMR International is a Thomson Reuters business that provides pharma industry metrics for drug discovery and development.

Combinatorial Chemistry It allows large numbers of compounds to be made by the systemic and repetitive covalent connection of a set of different 'building blocks' of varying structures to each other.

Compounding Bringing together excipient and solvent compo-nents into a homogeneous mix of active ingredients.

COPD Short for "chronic obstructive pulmonary disease," a disease characterized by reduced airflow and associated with symptoms such as breath and cough with sputum production.

CRO A CRO is a Contract Research Organization, i.e. an entity that provides technical and managerial support to the pharmaceutical companies in the form of research services outsourced on a contract basis.

CSO A CSO is a Contract Service Organization, i.e. any organization that provides pharmaceutical companies with a contract service. Among others, the term CSO includes:CROs (contract research organizations),CMOs (contract manufacturing organizations),SMOs (site management organizations).

Diagnostic A substance or group of substances used to identify a disease by analyzing cause and symptoms.

DNA DNA (Deoxyribonucleic Acid) represents the molecular basis for genes. Every inherited characteristic has its origin somewhere in the code of an organism's complement of DNA.

Efficacy The ability of a substance to produce a desired effect.

EFPIA The European Federation of Pharmaceutical Industries and Associations (EFPIA) is the association of the European research-based pharmaceutical industry.

ELF The European Lead Factory (ELF) is a public-private partnership that aims at accelerating early drug discovery in Europe.

EMEA EMEA is short for "European Agency for the Evaluation of Medicinal Products" or EMA (European Medicines Agency) that aims at harmonizing the work of national regulatory bodies in the European Union, and operates as a decentralized scientific agency responsible for the protection of public and animal health through the scientific evaluation and supervision of medicines.

Enzyme Macromolecules, mostly of protein nature, that function as (bio-) catalysts. They not only promote reactions but also function as regulators making sure the organism does not produce too much or too little of any chemical substance.

FDA Food and Drug Administration (U.S. regulatory approval body for new pharmaceutical products).

FIPnet FIPnet is an abbreviation for "fully integrated pharmaceutical network" and describes the open innovation concepts of the U.S. pharma company Eli Lilly.

Gene A natural unit of hereditary material that is the physical basis for the trans-mission of the charac-teristics of living organisms from one generation to another. The basic genetic material is essentially the same in all living organisms. It consists of deoxyribonucleic acid (DNA) in most organisms and ribonucleic acid (RNA) in certain viruses.

Gene Mapping Determination of the relative positions of genes on a DNA mole-cule (chromosome or plasmid) and of the distance (in linkage units or physical units) between them.

Gene Sequencing The determination of the sequence of bases in a DNA strand.

Gene Splicing The enzymatic attachment of one gene or part of a gene to another.

Generic Drug Replication of a prescription or non-prescription drug where the patent protection has expired. Generic drugs (also referred to as generics) are usually offered by firms that did not develop the drugs themselves but gained a license to sell the drug.

Genetic Diseases Diseases that occur because of a mutation in the genetic material.

Genetic Engineering The selective, deliberate alteration of genes by technological means.

Genetics The study of the genetic composition, heredity, and variation of organisms.

Genomic Library A collection of clones made from a set of randomly generated overlapping DNA fragments representing the entire genome of an organism.

Genomics The process of identifying genes involved in disease through the com-parison of the genomes of individuals with and without disease.

GERD Gastroesophageal reflux disease (GERD) is a long-term medical condition that affect 10-20% of the Western world population. Stomach contents come back up into the esophagus resulting in symptoms or complications, such as taste of acid in the back of the mouth, heartburn, bad breath, vomiting, or esophagitis. This is generally treated with proton pump inhibitors such as Pantoprazole.

HTS High-throughput Screening (HTS) refers to the process for rapid assessment of the activity of samples from a combinatorial library or other compound collection.

Immunology The study of how the body defends itself against disease.

IND Application An IND (Investigational New Drug) application is a document filed with the FDA prior to clinical trial of a new drug. It gives a full description of the new drug, such as where and how it is manufactured. An IND permission has to be kept active annually by sending, for example, annual reports. The IND is followed by the NDA (New Drug Application).

IP Intellectual property (IP) refers to creations of the intellect for which geographical monopolies in the form of patents, trademarks, utility models, industrial designs or copyrights can be designated.

iPSCs Induced pluripotent stem cells (iPSCs) are stem cells generated from adult cells (not from embryonic stem cells) that are used in biomedical research in the field of regenerative medicine.

Molecular Genetics The study of the nature and biochemistry of genetic material. It includes the technologies of genetic engineering.

NCE NCE is short for New Chemical Entity, referring to newly approved pharmaceutical products.

NDA NDA means New Drug Application, i.e. the process of determining the benefit-risk profile of a new drug after completion of the clinical tests and prior to approval for marketing.

NIH NIH is the U.S. National Institute of Health responsible for biomedical and public health research.

NME When a company submits new clinical investigation data for a therapeutic indication for the product which has an active moiety already approved by the FDA, it will be called a NME (New Molecular Entity). If it has no active moiety then it will be called a NCE.

NMR Nuclear magnatic resonance (NMR) spectroscopy is a technology that discovers the magnatic properties of atomic nuclei and thereby helps to determine the physical and chemical properties of molecules.

NPV The net present value (NPV) is an indicator of the profitability of an R&D investment that is calculated by subtracting the present values (PV) of cash outflows (including initial cost) from the present values of cash inflows over the time period for project initiation to end of product lifecycle.

NTDs Neglected tropical diseases (NTDs) are a group of infectious diseases (primarily HIV/AIDS, tuberculosis, and malaria) that are common in developing countries in Africa, Asia and Southern America.

Orphan Drug A drug that is believed to substantially increase the life expectancy of the treated patient for a particular disease. While developing an orphan drug, competitors are usually excluded from receiving a license to produce a similar drug for a finite period (usually 7 years), thereby allowing the company producing the drug to recuperate R&D expenses.

OTC Drug An OTC (Over-the-Counter) drug can usually be purchased without a prescription. An OTC drug is also sometimes referred to as a drug purely used for self-medication purposes and is typically used for minor ailments such as headache or the flu.

PDUFA PDFUA stands for "Prescription Drug User Fee Act" and describes a legal act in the U.S. that authorizes the FDA to collect fees from pharmaceutical companies that produce certain human drugs and biological products.

Pharmacodynamics Quantitative study of drug action.

Pharmacogenomics The study of how the response to a drug is affected by individual genetic variations. It is aimed at the prescription or development of drugs that maximize benefit and minimize side effects in individuals.

Pharmacokinetics Quantitative study of how drugs are taken up, biologically transformed, distributed, metabolized, and eliminated from the body.

PhRMA Pharmaceutical Research and Manufacturers of America (an organization representing the leading research-based pharmaceutical and biotechnology companies in the US).

PoC A clinical proof-of-concept (PoC) refers to a milestone in early clinical development (primarily phase II) that a new drug candidate has a useful amount of the desired clinical activity and that it can be tolerated when given to humans in the longer term. It is the decision for further investment and the starting point for full development (phase III) up to market authorization.

Proteomics The study of the entire protein output of cells. It refers to any protein-based approach that has the capacity to provide new information about proteins on a genome-wide scale.

PTRS PTRS stands for "probability of technical and regulatory success." It is an indicator for the success of R&D projects along the pharmaceutical value chain from drug discovery to preclinical and clinical development.

Recombinant Recombining of generic material from one species into alternate sequences.

RNA RNA (Ribonucleic Acid) is a single-strand molecule that partners with DNA to manufacture proteins.

ROI ROI is short for "return on investment" and describes the benefit to an investor resulting from the investment of financial resources.

SAR SAR is short for "structure-activity relationship", a relationship between the chemical structure of a molecule and its biologic activity. Analyzing the SAR helps to optimize the biologic effect of chemical molecules interacting with a drug target.

siRNA siRNA is short for "small interfering RNA", a class of double-stranded RNA of 20-25 base pairs that operates by interfering with the expression of certain genes. SiRNAs are used in biomedical research and drug development.

Target The target molecule is usually responsible for causing a respective disease. The targets of most drugs are proteins. The drug molecule, which is supposed to cure the respective disease, inserts itself into a functionally important crevice of the target protein, like a key in a lock. The drug molecule is then connected to the target and either induces or, more commonly, inhibits the protein's normal function.

Toxicogenomics Application of genetic and genomic methods to the study of toxicology.

Toxicology Study of poisonous substances in terms of their chemistry, effects, and treatments.

Treatment Investigational New Drug An Investigational New Drug that makes a promising new drug available to desperately ill patients as early in the drug development process as available. The FDA permits the drug to be used if there is preliminary evidence of efficacy and it treats a serious or life-threatening disease, or if there is no comparable therapy available.

Vaccine An agent containing antigens. It is used for stimulating the immune system of the recipient to produce specific antibodies providing active immunity and/or passive immunity in the progeny.

3GS 3GS is short for "third generation sequencing" or long-read sequencing, a further development of second generation sequencing (also referred to as NGS or next generation sequencing). Both sequencing methods are highly cost-efficient and effective technologies capable of producing large sequencing reads at exceptionally high coverages throughout the genome.

Bibliography

Agrafiotis, D. K., Alex, S., Dai, H., Derkinderen, A., Farnum, M., Gates, P., et al. (2007). Advanced biological and chemical discovery (ABCD): Centralizing discovery knowledge in an inherently decentralized world. *Journal of Chemical Informatics and Modeling, 47*, 1999–2014.

Agrafiotis, D. K., Lobanov, V. S., & Salemme, F. R. (2002). Combinatorial informatics in the post-genomics era. *Nature Reviews Drug Discovery, 1*, 337–346.

Arlington S., & Davies N. (2014). *Managing innovation in pharma*. PWC. https://www.pwc.com/gx/en/pharma-life-sciences/assets/pwc-managing-innovation-pharma.pdf. Accessed 25 April 2016.

Baumann, G. (2003). The challenge of innovation in the drug discovery process. Presentation at CTO-Roundtable 'Management of Pharmaceutical R&D in Turbulent Times – Perspectives and Trends'. Zurich 2003.

BCG. (2016). *Moving beyond the 'milkman' model in MedTech*. Boston: BCG Perspectives.

Beckman, R. A., Clark, J., & Chen, C. (2011). Integrating predictive biomarkers and classifiers into oncology clinical development programmes. *Nature Reviews Drug Discovery, 10*, 735–748.

Betz, U. (2011). Portfolio management in early stage drug discovery – a traveler's guide through uncharted territory. *Drug Discovery Today, 16*, 609–618.

Bhattacharyya, O., Wu, D., Mossman, K., Hayden, L. Gill, P., Cheng, Y.-L., et al. (2017). Criteria to assess potential reverse innovations: Opportunities for shared learning between high and low income countries. *Globalization and Health, 13*(4), 1-8.

Bleicher, K. H., Böhm, H. J., Müller, K., & Alanine, A. I. (2003). A guide to drug discovery: Hit and lead generation: Beyond high-throughput screening. *Nature Reviews Drug Discovery, 2*, 369–378.

Bode-Greuel, K., & Greuel, J. M. (2004). Determining the value of drug development candidates and technology platforms. *Journal of Commercial Biotechnology, 11*, 155–170.

Booz Allen & Hamilton. (1997). *In vivo, making combinatory chemistry pay*. New York: Booz Allen & Hamilton Report.

Boutellier, R., Gassmann, O., & von Zedtwitz, M. (2008). *Managing global innovation: Uncovering the secrets of future competitiveness* (3rd ed.). Berlin: Springer.

Butcher, E. C. (2005). Can cell systems biology rescue drug discovery? *Nature Reviews Drug Discovery, 4*, 461–467.

Chakma, J., Calcagno, J. L., Behbahani, A., & Mojtahedian, S. (2009). Is it virtuous to be virtual? The VC viewpoint. *Nature Biotechnology, 27*(10), 886–888.

Chesbrough, H. (2003). *Open innovation: The new imperative for creating and profiting from technology*. Boston, MA: Harvard Business School Press.

Comanor, W. S., & Scherer, F. M. (2013). Mergers and innovation in the pharmaceutical industry. *Journal of Health Economics, 32*(1), 106–113.

Corsi, S., & von Zedtwitz, M. (2016). Reverse innovation – a new world order for global innovation? *The European Business Review, 2016*, 73–77.

Courtois, Y., McPhee, D., & Rerolle J.F. (2012). *Profitability and royalty rates across industries: Some preliminary evidence.* KPMG International. https://www.kpmg.com/Global/en/IssuesAndInsights/ArticlesPublications/Documents/gvi-profitability-v6.pdf. Accessed 27 April 2016.

Crisp, N. (2010). *Turning the world upside down: The search for global health in the 21st century.* London: Royal Society of Medicine Press.

Damodaran, A. (2016). *Datasets.* http://people.stern.nyu.edu/adamodar/NewHomePage/data.html. Accessed 12 December 2016.

Davis, S., & Botkin, J. (1994). The coming of knowledge-based business. *Harvard Business Review, 5,* 165–170.

Deloitte. (2011). *The next phase: Opportunities in China's pharmaceutical market.* National Industry Program Report.

Deloitte. (2014). *2015 Global life science outlook. Adapting in an era of transformation.* https://www2.deloitte.com/content/dam/Deloitte/global/Documents/Life-Sciences-Health-Care/gx-lshc-2015-life-sciences-report.pdf. Accessed 25 April 2016.

DePasse, J., & Lee, P. (2013). A model for 'reverse innovation' in health care. *Globalization and Health, 9,* 40.

Dierks, A., Kuklinski, C. P. J., & Moser, R. (2013). How institutional change reconfigures successful value chains: The case of western pharma corporations in China. *Thunderbird International Business Review, 55*(2), 153–171.

Dorsch, H., Jurock, A. E., Schoepe, S., Lessl, M., & Asadullah, K. (2015). Grants4Targets—an open innovation initiative to foster drug discovery collaborations between academia and the pharmaceutical industry. *Nature Reviews Drug Discovery, 14,* 74–76.

Eder, J., Sedrani, R., & Wiesmann, C. (2014). The discovery of first-in-class drugs: Origins and evolution. *Nature Reviews Drug Discovery, 13,* 577–587.

European Commission. (2014). *The 2014 EU Industrial R&D investment scoreboard.* European Commission—Joint Research Centre. http://iri.jrc.ec.europa.eu/scoreboard.html. Accessed 8 June 2016.

European Commission. (2015). *The 2015 EU Industrial R&D investment scoreboard.* European Commission—Joint Research Centre. http://iri.jrc.ec.europa.eu/scoreboard15. Accessed 8 June 2016.

Evaluate. (2015). *World preview 2015, outlook to 2020.* http://info.evaluategroup.com/rs/607-YGS-364/images/wp15.pdf. Accessed 15 February 2016.

Evaluate. (2016). *World Preview 2016, Outlook to 2022.* 9th Edition, September 2016.

Export. (2016). *China Pharmaceuticals.* Online at https://www.export.gov/article?id=China-Pharmaceuticals

FDA. (2014). *Guidance for industry and FDA staff qualification process for drug development tools.* Silver Spring: FDA.

Festel, G., & Polastro, E. (2002). Dritte arbeiten häufig kostengünstiger. *Chemische Rundschau, 55* (13), 7.

Flanagan, M. (2015). *2015 In review: Pharma's largest in-licensing deals – cancer continues to carry the day.* FirstWord Pharma, http://www.firstwordpharma.com/node/1339317?tsid=33#axzz40sjGtMpV. Accessed 27 April 2016.

Gassmann, O., & Keupp, M. M. (2007). The competitive advantage of early and rapidly internationalising SMEs in the biotechnology industry: A knowledge based view. *Journal of World Business, 42,* 350–366.

Gassmann, O., & von Zedtwitz, M. (2003a). Trends and determinants of managing virtual R&D teams. *R&D Management, 33*(3), 243–262.

Gassmann, O., & von Zedtwitz, M. (2003b). Innovation processes in transnational corporations. In L. Shavinina (Ed.), *The international handbook on innovation* (pp. 702–714). Oxford: Pergamon.

Gaze, L. & Breen, J. (2012). *The economic power of orphan drugs.* Thomson Reuters at http://thomsonreuters.com/content/dam/openweb/documents/pdf/pharma-life-sciences/white-paper/1001450.pdf. Accessed 26 May 2016.

Govindarajan, V., & Ramamurti, R. (2011). Reverse innovation, emerging markets, and global strategy. *Global Strategy Journal, 1*(3–4), 191–205.

Grabowski, H., Vernon, J., & DiMasi, J. A. (2002). Returns on research and development for 1990s new drug introductions. *Pharmacoeconomics, 20,* 11–29.

Greis, N. P., Dibner, M. D., & Bean, A. S. (1995). External partnering as a response to innovation barriers and global competition in biotechnology. *Research Policy, 24,* 609–630.

Grimes, S., & Miozzo, M. (2015). Big pharma's internationalization of R&D to China. *European Planning Studies, 23*(9), 1873–1894.

Guo, Y. (2016). *TCM going global.* China.org.cn. Accessed 11 March 2016.

Hadengue, M., de Marcellis-Warin, N., von Zedtwitz, M., & Warin, T. (2017). Avoiding the pitfalls of reverse innovation. *Research Technology Management, 60*(3), 40–47.

Hajduk, P. J., & Greer, J. (2007). A decade of fragment-based drug design: Strategic advances and lessons learned. *Nature Reviews Drug Discovery, 6,* 211–219.

Harris, M., Weisberger, E., Silver, D., & Macinko, J. (2015). 'They hear "Africa" and they think that there can't be any good services'–perceived context in cross-national learning: A qualitative study of the barriers to reverse innovation. *Globalization and Health, 11,* 45.

HHS & FDA. (2005). *Drug-diagnostic co-development concept paper.* Department of Health and Human Services.

Hofmann, D. (1997). Das virtuelle Unternehmen. *Neue Zürcher Zeitung,* p. 29. Accessed 25 Oct 1997.

Hurko, O., & Jones, G. K. (2011). Valuation of biomarkers. *Nature Reviews Drug Discovery, 10,* 253–254.

IMS. (2016). *Price declines after branded medicines lose exclusivity in the U.S.* IMS Institute for Healthcare Informatics, January 2016.

ITA. (2016). *China – 2016 top markets report pharmaceuticals country case study.* Washington, DC: International Trade Association, D.O.C.

Karst, K. R. (2017). Orphan drug approvals and designations dipped in 2016, but orphan drug designation requests skyrocketed. Online at http://www.fdalawblog.net/fda_law_blog_hyman_phelpssince25Jan2017. Accessed 22 February 2017.

Kermani, F. Z. (2014). Drug discovery partnerships between UK CROs and the Swiss pharma sector. *Pharmaceutical Technology Europe, 26,* 8–11.

Kinch, M. S. (2015). An overview of FDA-approved biologics medicines. *Drug Discovery Today, 20,* 393–398.

Kitano, H. (2002). Computational systems biology. *Nature, 420,* 206–210.

KnowledgeWharton. (2013, September). China's bitter medicine for foreign drug companies. KnowledgeWharton. Available at http://knowledge.wharton.upenn.edu/article/chinas-bitter-medicine-foreign-drug-companies/

Kober, S. (2008). The evolution of specialty pharmacy. *Biotechnology Healthcare, 5*(2), 50–51.

Kogej, T., Blomberg, N., Greasley, P. J., Mundt, S., Vainio, M. J., Schamberger, J., et al. (2013). Big pharma screening collections: More of the same or unique libraries? The AstraZeneca–Bayer pharma AG case. *Drug Discovery Today, 18,* 1014–1024.

Kollmer, H., & Dowling, M. (2004). Licensing as a commercialisation strategy for new technology-based firms. *Research Policy, 33,* 1141–1151.

Kramer, R., & Coher, D. (2004). Functional genomics to new drug targets. *Nature Reviews Drug Discovery, 3,* 965–972.

Leutenegger, J. -M. (1994). Wettbewerbsorientierte Informationssysteme in der Schweizer Pharma-Branche.

Lin, B.-W. (2001). *Strategic alliances and innovation networks in the biopharmaceutical industry.* Hsinchu: Institute of Technology Management, National Tsinghua University.

Little, A. D. & Solvias. (2002, June). External synthesis services for research and development in the pharmaceutical industry. *Market Study.*

Liu, C. X., & Xiao, P. G. (2002). Recalling the research and development of new drugs originating from Chinese traditional and herbal drugs. *Asian Journal of Drug Metabolism and Pharmacokinetics, 2* (2), 133–156.

Macarron, R. (2006). Critical review of the role of HTS in drug discovery. *Drug Discovery Today, 11*, 277–279.

Macarron, R., Banks, M. N., Bojanic, D., Burns, D. J., Cirovic, D. A., Garyantes, T., et al. (2011). Impact of high-throughput screening in biomedical research. *Nature Reviews Drug Discovery, 10*, 188–195.

Manji, H. K., Inel, T. R., & Narayan, V. A. (2014). Harnessing the informatics revolution for neuroscience drug R&D. *Nature Reviews Drug Discovery, 13*, 561–562.

Maznevski, M. S., & Chudoba, K. M. (2000). Bridging space over time: Global virtual team dynamics and effectiveness. *Organizational Science, 11*(5), 473–492.

McKinsey. (2012, February). Innovating in China's pharma market: An interview with AstraZeneca's head of R&D in Asia and emerging markets. *McKinsey Insights*.

Megantz, R. C. (2002). *Technology management – developing and implementing effective licensing programs*. New York: Wiley.

Mohamed, S., & Syed, B. A. (2013). Commercial prospects for genomic sequencing technologies. *Nature Reviews Drug Discovery, 12*, 341–342.

Mullard, A. (2016). 2015 FDA drug approvals. *Nature Reviews Drug Discovery, 15*, 73–76.

Newman, D. J., Cragg, G. M., & Snader, K. M. (2003). Natural products as sources of new drugs over the period 1981–2002. *Journal of Natural Products, 66*, 1022–1037.

Nicholls, A. & Brayshaw, L. (2014). World industry outlook: Healthcare and pharmaceuticals. *The Economist Intelligence Unit*. http://pages.eiu.com/rs/eiu2/images/GlobalOutlook_Healthcare.pdf. Accessed 25 April 2016.

Nicholson, J. K., Connelly, J., Lindon, J. C., & Holmes, E. (2002). Metabonomics: A platform for studying drug toxicity and gene function. *Nature Reviews Drug Discovery, 1*, 153–161.

Nightingale, P. (2000). Economies of scale in experimentation: Knowledge and technology in pharmaceutical R&D. *Industrial and Corporate Change, 9*(2), 315–359.

Nonaka, I., & Takeuchi, H. (1995). *The knowledge-creating company. How Japanese companies create the dynamics of innovation*. New York: Oxford.

Paolini, G. V., Shapland, R. H. B., Hoorn van, W. P., Mason, J. S., & Hopins, A. L. (2006). Global mapping of pharmacological space. *Nature Biotechnology, 24*, 805–815.

Paul, S. M., Mytelka, D. S., Dunwiddie, C. T., Persinger, C. C., Munos, B. H., Lindborg, S. R., et al. (2010). How to improve R&D productivity: The pharmaceutical industry's grand challenge. *Nature Reviews Drug Discovery, 9*, 203–214.

PharmaSource. (2017). http://www.pharmsource.com/market/how-big-is-the-market-for/

Philipp, J. (2014, December 8). Beware of fake prescription drugs smuggled from China. *The Epoch Times*.

Pharma Information. (2002). *Swiss Health Care and pharmaceutical market. Edition 2002*. Basel: Interpharma.

PhRMA. (2015). *2015 Profile biopharmaceutical research industry*. PhRMA website at http://www.phrma.org/sites/default/files/pdf/2015_phrma_profile.pdf. Accessed 20 June 2016.

Plenge, R. M., Scolnick, E. M., & Altshuler, D. (2013). Validating therapeutic targets through human genetics. *Nature Reviews Drug Discovery, 12*, 581–594.

Porter, M. E. (1985). *Competitive advantage: Creating and sustaining superior performance*. New York: Macmillan.

Prud'homme, D. (2016). Forecasting threats and opportunities for foreign innovators in China's strategic emerging industries: A policy-based analysis. *Thunderbird International Business Review, 58*, 103–115.

Rask-Andersen, M., Sällman Almén, M., & Schiöth, H. B. (2011). Trends in the exploitation of novel drug targets. *Nature Reviews Drug Discovery, 10*, 579–590.

Recombinant Capital. (2005). Analyst's notebook. *Trends*. http://www.recap.com/consulting.nsf/ANB_tab_trends?openform. Accessed 11 February 2005.

Reuters. (2002). *Pharmaceutical innovation – an analysis of leading companies and strategies*. Reuters Business Insight, Healthcare.

Reuters. (2003a). *The blockbuster drug outlook to 2007: Identifying, creating and maintaining the pharmaceutical industry's growth drivers*. Reuters Business Insight, Healthcare.

Reuters. (2003b). *Patent protection strategies: Maximizing market exclusivity*. Reuters Business Insight, Healthcare.

Reuters. (2003c). Pharmaceutical R&D outsourcing strategies – an analysis of market drivers and resistors to 2010. Reuters Business Insight, Healthcare.

Robbins-Roth, C. (2001). Zukunftsbranche Biotechnologie. Gabler.

Saftlas, H. (2001). *Industry surveys, healthcare: Pharmaceuticals* (p. 32). New York: Standard & Poors.

Scannell, J. W., Blanckley, A., Boldon, H., & Warrington, B. (2012). Diagnosing the decline in pharmaceutical R&D efficiency. *Nature Reviews Drug Discovery, 11*, 191–200.

Schuhmacher, A., Gassmann, O., & Hinder, M. (2016). Changing R&D models in research-based pharmaceutical companies. *Journal of Translational Medicine, 14*, 105–115.

Searls, D. B. (2003). Pharmacophylogenomics: Genes, evolution and drug targets. *Nature Reviews Drug Discovery, 2*, 613–623.

Shi, Y.-Z., Hu, H., & Wang, C. (2014). Contract research organizations (CROs) in China: Integrating Chinese research and development capabilities for global drug innovation. *Globalization and Health, 10*, 78.

Snowdon, A. W., Bassi, H., Scarffe, A. D., & Smith, A. D. (2015). Reverse innovation: An opportunity for strengthening health systems. *Globalization and Health, 11*, 2.

Speedel. (2004). Introducing the Speedel group. Company presentation. http://www.speedelgroup.com. Accessed 10 September 2004.

Swinney, D. C., & Anthony, J. (2011). How were new medicines discovered? *Nature Reviews Drug Discovery, 10*, 507–519.

Swissmedic. (2002). http://www.swissmedic.ch. Accessed September 2002.

Syed, S. B., Dadwal, V., & Martin, G. (2013). Reverse innovation in global health systems: Towards global innovation flow. *Globalization and Health, 9*, 36.

Thomas, K. & Pollack, A. (2015, 15 July). Specialty pharmacies proliferate, along with questions. *New York Times*. Retrieved 5 October 2015.

Ulrich, R., & Friend, S. H. (2002). Toxicogenomics and drug discovery: Will new technologies help us produce better drugs? *Nature Reviews Drug Discovery, 1*, 84–88.

van der Greef, J., & McBurney, N. (2005). Rescuing drug discovery: In vivo systems pathology and systems pharmacology. *Nature Reviews Drug Discovery, 4*, 961–967.

von Zedtwitz, M., Corsi, S., Soberg, P., & Frega, R. (2015). A typology of reverse innovation. *Journal of Product Innovation Management, 32*(1), 12–28.

von Zedtwitz, M., & Gassmann, O. (2002). Market versus technology drive in R&D internationalization: Four different patterns of managing research and development. *Research Policy, 31*(4), 569–588.

von Zedtwitz, M., & Gassmann, O. (2016). Global corporate RnD to and from emerging countries. In S. Dutta, B. Lanvin, & S. Wunsch-Vincent (Eds.), *The global innovation index 2016* (pp. 125–131). Ithaca, NY: Johnson Cornell University.

Waldmeir, P. (2015, 11 November). China looks to traditional medicine as tonic to boost growth. *Financial Times*.

Wall Street Journal. (2015). *Why the U.S. pays more than other countries for drugs*. Article online at https://www.wsj.com/articles/why-the-u-s-pays-more-than-other-countries-for-drugs-1448939481

Wang, A., & von Zedtwitz, M. (2005). Developing the pharmaceutical business in China – the case of Novartis. In G. Festel, A. Kreimeyer, U. Oels, & M. von Zedtwitz (Eds.), *The chemical and pharmaceutical industry in China – challenges and threats for foreign companies* (pp. 109–119). Heidelberg: Springer.

Webber, D. E. (2005). China's approach to innovative pharmaceutical R&D: A review. In G. Festel, A. Kreimeyer, U. Oels, & M. von Zedtwitz (Eds.), *The chemical and pharmaceutical industry in China – challenges and threats for foreign companies* (pp. 121–131). Heidelberg: Springer.

Wenk, M. R. (2005). The emerging field of lipidomics. *Nature Reviews Drug Discovery, 4,* 594–610.

Weyrauch, T., & Herstatt, C. (2016). What is frugal innovation? Three defining criteria. *Journal for Reverse Innovation, 2*(1).

Whittaker, E., & Bower, D. J. (1994). A shift to external alliances for product development in the pharmaceutical industry. *R&D Management, 24*(3), 249–260.

Windhover. (2000). *Opportunism knocks. Windhover's review of emerging medical ventures.* Vol. 5, No. 4, p. 32. http://www.windhover.com/contents/monthly/exex/e_2000900064.htm. Accessed 19 December 2004.

Windhover. (2003). *In-licensing: Still a difficult model. Windhover's review of emerging medical ventures.* Vol. 8, No. 10. http://www.windhover.com/contents/monthly/exex/e_2003900172.htm. Accessed 19 December 2004.

Zeller, C. (2001). *Globalisierungsstrategien – der Weg von Novartis.* Heidelberg: Springer.

Zeschky, M., Winterhalter, S., & Gassmann, O. (2014). From cost to frugal and reverse innovation: Mapping the field and implications for global competitiveness. *Research Technology Management, 57*(4), 1–8.

Index

© Springer International Publishing AG, part of Springer Nature 2018   177
O. Gassmann et al., *Leading Pharmaceutical Innovation*,
https://doi.org/10.1007/978-3-319-66833-8

Printed in the United States
By Bookmasters